A_PEX Pre-Calculus
Version 1.0

Amy Givler Chapman, Ph.D.

Department of Applied Mathematics

Virginia Military Institute

Contributing Authors

Meagan Herald, Ph.D.

Department of Applied Mathematics

Virginia Military Institute

Jessica Libertini, Ph.D.

Department of Applied Mathematics

Virginia Military Institute

Preface

A Note on Using this Text

Thank you for taking your time to read this preface. We will briefly share some key features of this text to (hopefully) improve your experience using it.

For Instructors: How to Use this Text

This text was written as a prequel to the A$_{\mathrm{E}}^{\mathrm{P}}$XCalculus series, a three-volume series on Calculus. This text is not intended to fully prepare students with all of the mathematical knowledge they need to tackle Calculus, rather it is designed to review mathematical concepts that are often stumbling blocks in the Calculus sequence. It starts basic and builds to more complex topics. This text is written so that each section and topic largely stands on its own, making it a good resource for students in Calculus who are struggling with the supporting mathematics found in Calculus courses. The topics were chosen based on experience; several instructors in the Applied Mathematics Department at the Virginia Military Institute (VMI) compiled a list of topics that Calculus students commonly struggle with, giving the focus of this text. This allows for a more focused approach; at first glance one of the obvious differences from a standard Pre-Calculus text is its size.

This text, as well as the three volumes of the A$_{\mathrm{E}}^{\mathrm{P}}$XCalculus series, is available separately for free at www.apexcalculus.com. All four texts can be purchased as bound volumes for $15 or less per text at Amazon.com.

For Students: How to Read this Text

Many mathematical texts are written in very formal, succinct language. This is a terrific approach if this text is simply used as a resource for someone who is already comfortable with the ideas in the text, but can make it difficult for anyone who is new to the material. This text was written in a different fashion. Its goal is to show you mathematical ideas and concepts explained in an informal style so that your focus is on learning the math, not trying to decipher the sentences.

This text is written with many examples. Each new idea is shown through several examples, starting with a straightforward example and working up to more complex examples of the idea. These examples, and the exercises in the text, come from the mathematics as it appears in Calculus. You may notice that if you use this text as a resource while you are taking Calculus that many of the problems in the text come from problems in Calculus. For example, many of the questions asking you to simplify a function are really unsimplified derivatives of common function types, a type of function with a special meaning that is used in Calculus.

Additionally, the later sections of this text will reinforce many of the ideas of the earlier sections. This is entirely on purpose. In Calculus, you will need to use many of these skills in the solution of a larger problem. The larger problems almost never tell you the names of the skills you will use. This means that you need to identify which skills to use and when to use them. To help get you used to these types of problems, this text often gives you problems that require skills from earlier sections, without telling you about it.

Finally, answers (but not solutions) to all exercises are provided in an appendix at the end of the text. We highly recommend checking your answer to each exercise before moving on to another exercise. This will prevent you from practicing skills incorrectly and will save you the potential frustration of finding out that you have done several problems and made the same mistake on all of them.

Thanks

Many people contributed to this text, in ways small and large. First, thanks are due to the VMI students who first used this text during the Summer Transition Program (STP) in 2017 and the VMI cadets who used it during Fall 2017. These students were diligent readers who found many typos and made suggestions that greatly improved the usefulness of the text for students. Second, thanks to Meagan Herald, not only for proofreading the entire text and answer key multiple times, but for never complaining about using a work in progress for the basis of VMI's Pre-Calculus course. Major revisions were made between STP 2017 and Fall 2017 with her help and guidance, including reordering the sections of the text so that exercises and examples did not use concepts that had not been discussed by that point. Meagan also initially developed a list of topics that formed the basis for those used in this text and well as the Pre-Calculus course at VMI.

Additionally, I would like to thank Jessica Libertini for motivating me to complete this text and providing me with many of the problems used in the text as well as pedagogical advice. Furthermore, the hard work of Greg Hartman who authored and maintains the AP_{E}XCalculus series needs to be acknowledged; he is responsible for creating the formatting files that hold this text together. Throughout this process, we were also given large amounts of support by the

Notes:

Applied Mathematics Department at the Virginia Military Institute, most notably the department head, Troy Siemers who supported our efforts from the beginning.

Finally, thanks are due to my husband Jonathan Chapman who supported me as I worked on this text extra hours outside of the office and who provided me with technical support to streamline the creation process.

APEX – Affordable Print and Electronic teXts

APEX is a consortium of authors who collaborate to produce high-quality, low-cost textbooks. The current textbook-writing paradigm is facing a potential revolution as desktop publishing and electronic formats increase in popularity. However, writing a good textbook is no easy task, as the time requirements alone are substantial. It takes countless hours of work to produce text, write examples and exercises, edit and publish. Through collaboration, however, the cost to any individual can be lessened, allowing us to create texts that we freely distribute electronically and sell in printed form for an incredibly low cost.

Each text is available as a free .pdf, protected by a Creative Commons Attribution - Noncommercial 4.0 copyright. That means you can give the .pdf to anyone you like, print it in any form you like, and even edit the original content and redistribute it. If you do the latter, you must clearly reference this work and you cannot sell your edited work for money.

We encourage others to adapt this work to fit their own needs. One might add sections that are "missing" or remove sections that your students won't need. The source files can be found at github.com/APEXCalculus.

You can learn more at www.vmi.edu/APEX.

Contents

1: Numbers and Functions

When we first start learning about numbers, we start with the counting numbers: 1, 2, 3, etc. As we progress, we add in 0 as well as negative numbers and then fractions and non-repeating decimals. Together, all of these numbers give us the set of *real numbers*, denoted by mathematicians as $\mathbb{R}$, numbers that we can associate with concepts in the real world. These real numbers follow a set of rules that allow us to combine them in certain ways and get an unambiguous answer. Without these rules, it would be impossible to definitively answer many questions about the world that surrounds us.

In this chapter, we will discuss these rules and how they interact. We will see how we can develop our own "rules" that we call functions. In calculus, you will be manipulating functions to answer application questions such as optimizing the volume of a soda can while minimizing the material used to make it or computing the volume and mass of a small caliber projectile from an engineering drawing. However, in order to answer these complicated questions, we first need to master the basic set of rules that mathematicians use to manipulate numbers and functions.

Additionally, we will learn about some special types of functions: logarithmic functions and exponential functions. Logarithmic functions and exponential functions are used in many places in calculus and differential equations. Logarithmic functions are used in many measurement scales such as the Richter scale that measures the strength of an earthquake and are even used to measure the loudness of sound in decibels. Exponential functions are used to describe growth rates, whether it's the number of animals living in an area or the amount of money in your retirement fund. Because of the varied applications you will see in calculus, familiarity with these functions is a must.

1.1   Real Numbers

We begin our study of *real numbers* by discussing the rules for working with these numbers and combining them in a variety of ways. In elementary school, we typically start by learning basic ways of combining numbers, such as addition, subtraction, multiplication, and division, and later more advanced operations like exponents and roots. We will not be reviewing each of these operations, but we will discuss how these operations interact with each other and how to determine which operations need to be completed first in complicated mathematical expressions.

You are probably already familiar with the phrase "order of operations." When mathematicians refer to the order of operations they are referring to a guideline for which operations need to be computed first in complicated expressions:

1. Parentheses

2. Exponents

3. Multiplication/Division

4. Addition/Subtraction

Often we learn phrases such as "Please Excuse My Dear Aunt Sally" to help remember the order of these operations, but this guideline glosses over a few important details. Let's take a look at each of the operations in more detail.

Parentheses

There are two important details to focus on with parentheses: nesting and "implied parentheses." Let's take a look at an example of nested parentheses first:

Example 1   **Nested Parentheses**
Evaluate

$$2 \times (3 + (4 \times 2)). \tag{1.1}$$

Solution   Here we see a set of parentheses "nested" inside of a second set of parentheses. When we see this, we want to start with the inside set of parentheses first:

$$2 \times (3 + (4 \times 2)) = 2 \times (3 + (8)) \tag{1.2}$$

Once we simplify the inside set of parentheses to where they contain only a single number, we can drop them. Then, it's time to start on the next parentheses layer:

$$\begin{aligned} 2 \times (3 + (8)) &= 2 \times (3 + 8) \\ &= 2 \times (11) \\ &= 22 \end{aligned} \tag{1.3}$$

This gives our final answer:

$$\boxed{2 \times (3 + (4 \times 2)) = 22}$$

Sometimes this can get confusing when we have lots of layers of parentheses. Often, you will see mathematicians use both parentheses, "(" and ")", and

Notes:

brackets, "[" and "]". This can make it a bit easier to see where the parentheses/brackets start and where they end. Let's look at an example:

Example 2   **Alternating Parentheses and Brackets**

Evaluate

$$(2 + (3 \times (4 + (2 - 1)) - 1)) + 2. \qquad (1.4)$$

Solution

$$
\begin{aligned}
(2 + (3 \times (4 + (2 - 1)) - 1)) + 2 &= [2 + (3 \times [4 + (2 - 1)] - 1)] + 2 \\
&= [2 + (3 \times [4 + (1)] - 1)] + 2 \\
&= [2 + (3 \times [4 + 1] - 1)] + 2 \\
&= [2 + (3 \times [5] - 1)] + 2 \\
&= [2 + (3 \times 5 - 1)] + 2 \\
&= [2 + (15 - 1)] + 2 \\
&= [2 + (14)] + 2 \\
&= [2 + 14] + 2 \\
&= [16] + 2 \\
&= 16 + 2 = 18
\end{aligned}
\qquad (1.5)
$$

We started by finding the very inside set of parentheses: $(2 - 1)$. The next layer of parentheses we changed to brackets: $[4 + (2 - 1)]$. We continued alternating between parentheses and brackets until we had found all layers. As before, we started with evaluating the inside parentheses first: $(2 - 1) = (1) = 1$. The next layer was brackets: $[4 + 1] = [5] = 5$. Next, we had more parentheses: $(3 \times 5 - 1) = (15 - 1) = (14) = 14$. Then, we had our final layer: $[2 + 14] + 2 = [16] + 2 = 16 + 2 = 18$.

This gives our final answer:

$$\boxed{(2 + (3 \times (4 + (2 - 1)) - 1)) + 2 = 18}$$

When you are working these types of problems by hand, you can also make the parentheses bigger as you move out from the center:

$$(2 + (3 \times (4 + (2 - 1)) - 1)) + 2 = \left[2 + \left(3 \times \left[4 + (2 - 1)\right] - 1\right)\right] + 2$$

This may make it easier to see which parentheses/brackets are paired. You never have to switch a problem from all parentheses to parentheses and brackets, but you can alternate between them as you please, as long as you match parentheses "(" with parentheses ")" and brackets "[" with brackets "]".

Notes:

There's one more thing that we have to be careful about with parentheses, and that is "implied" parenthesis. Implied parentheses are an easy way to run into trouble, particularly if you are using a calculator to help you evaluate an expression. So what are implied parentheses? They are parentheses that aren't necessarily written down, but are implied. For example, in a fraction, there is a set of implied parentheses around the numerator and a set of implied parentheses around the denominator:

$$\frac{3+4}{2+5} = \frac{(3+4)}{(2+5)} \tag{1.6}$$

You will almost never see the second form written down, however the first form can you get into trouble if you are using a calculator. If you enter $3+4\div2+5$ on a calculator, it will first do the division and then the two additions since it can only follow the order of operations (listed earlier). This would give you an answer of 10. However, the work to find the actual answer is shown below.

Example 3   **Implied Parentheses in a Fraction**
Evaluate the expression in (1.6).

Solution   First, let's go back and find (1.6). You may have noticed that the fractions above have (1.6) next to them on the right side of the page. This tells us that (1.6) is referring to this expression. Now that we know what we are looking at, let's evaluate it:

$$\frac{3+4}{2+5} = \frac{(3+4)}{(2+5)}$$
$$= \frac{(7)}{(7)} \tag{1.7}$$
$$= \frac{7}{7}$$
$$= 1$$

This reflects what we would get on a calculator if we entered $(3+4) \div (2+5)$, giving us our final answer:

$$\boxed{\frac{3+4}{2+5} = 1}$$

As you can see, leaving off the implied parentheses drastically changes our answer. Another place we can have implied parentheses is under root operations, like square roots:

Notes:

Example 4   **Implied Parenthesis Under a Square Root**

Evaluate

$$\sqrt{12 \times 3} - 20$$

.

Solution

$$\begin{aligned} \sqrt{12 \times 3} - 20 &= \sqrt{(12 \times 3)} - 20 \\ &= \sqrt{(36)} - 20 \\ &= 6 - 20 \\ &= -14 \end{aligned} \tag{1.8}$$

This gives our final answer:

$$\boxed{\sqrt{12 \times 3} - 20 = -14}$$

Most calculators will display $\sqrt{(}$ when you press the square root button; notice that this gives you the opening parenthesis, but not the closing parenthesis. Be sure to put the closing parenthesis in the correct spot. If you leave it off, the calculator assumes that everything after $\sqrt{(}$ is under the root otherwise. This also applies to other kinds of roots, like cube roots. In the expression in Example 4, without a closing parenthesis, a calculator would give us $\sqrt{(}12 \times 3 - 20 = \sqrt{(}36 - 20 = \sqrt{(}16 = 4$.

We'll see another example of a common issue with implied parentheses in the next section.

Exponents

With exponents, we have to be careful to only apply the exponent to the term immediately before it.

Example 5   **Applying an Exponent**

Evaluate

$$2 + 3^3$$

.

Solution

$$\begin{aligned} 2 + 3^3 &= 2 + 27 \\ &= 29 \end{aligned} \tag{1.9}$$

Notes:

Notice we only cubed the 3 and not the expression $2 + 3$, giving us a final answer of

$$2 + 3^3 = 29$$

This looks relatively straight-forward, but there's a special case where it's easy to get confused, and it relates to implied parentheses.

Example 6   **Applying an Exponent When there is a Negative**
Evaluate

$$-4^2$$

.

Solution

$$
\begin{aligned}
-4^2 &= -(4^2) \\
&= -(16) \\
&= -16
\end{aligned}
\tag{1.10}
$$

Here, our final answer is

$$-4^2 = -16$$

Notice where we placed the implied parenthesis in the problem. Since exponents only apply to the term immediately before them, only the 4 is squared, not -4. Taking the extra step to include these implied parentheses will help reinforce this concept for you; it forces you to make a clear choice to show how the exponent is being applied. If we wanted to square -4, we would write $(-4)^2$ instead of -4^2.

Note that we don't take this to the extreme; 12^2 still means "take 12 and square it," rather than $1 \times (2^2)$.

It's also important to note that all root operations, like square roots, count as exponents, and should be done after parentheses but before multiplication and division.

Multiplication and Division

In our original list for the order of operations, we listed multiplication and division on the same line. This is because mathematicians consider multiplication

Notes:

and division to be on the same level, meaning that one does not take precedence over the other. This means you should not do *all* multiplication steps and then *all* division steps. Instead, you should do multiplication/division from left to right.

Example 7   **Multiplication/Division: Left to Right**

Evaluate

$$6 \div 2 \times 3 + 1 \times 8 \div 4$$

Solution

$$\begin{aligned} 6 \div 2 \times 3 + 1 \times 8 \div 4 &= 3 \times 3 + 1 \times 8 \div 4 \\ &= 9 + 1 \times 8 \div 4 \\ &= 9 + 8 \div 4 \\ &= 9 + 2 \\ &= 11 \end{aligned} \qquad (1.11)$$

Since this expression doesn't have any parentheses or exponents, we look for multiplication or division, starting on the left. First, we find $6 \div 2$, which gives 3. Next, we have 3×3, giving 9. The next operation is an addition, so we skip it until we have no more multiplication or division. That means that we have $1 \times 8 = 8$ next. Our last step at this level is $8 \div 4 = 2$. Now, we only have addition left: $9 + 2 = 11$. Our final answer is

$$\boxed{6 \div 2 \times 3 + 1 \times 8 \div 4 = 11}$$

Note that we get a different, incorrect, answer of 3 if we do all the multiplication first and then all the division.

Addition and Subtraction

Just like with multiplication and division, addition and subtraction are on the same level and should be performed from left to right:

Example 8   **Addition/Subtraction: Left to Right**

Evaluate

$$1 - 3 + 6$$

Notes:

Solution

$$1 - 3 + 6 = -2 + 6$$
$$= 4$$

(1.12)

By doing addition and subtraction on the same level, from left to right, we get a final answer of

$$\boxed{1 - 3 + 6 = 4}$$

Again, note that if we do all the addition and then all the subtraction, we get an incorrect answer of -8.

Summary: Order of Operations

Now that we've refined some of the ideas about the order of operations, let's summarize what we have:

1. Parentheses (including implied parentheses)

2. Exponents

3. Multiplication/Division (left to right)

4. Addition/Subtraction (left to right)

Let's walk through one example that uses all of our rules.

Example 9   **Order of Operations**
Evaluate

$$-2^2 + \sqrt{6 - 2} - 2(8 \div 2 \times (1 + 1))$$

Solution   Since this is more complicated than our earlier examples, let's make a table showing each step on the left, with an explanation on the right:

Notes:

$-2^2 + \sqrt{6-2} - 2(8 \div 2 \times (1+1)) =$	We have a bit of everything here, so let's write down any implied parentheses first.
$= -2^2 + \sqrt{(6-2)} - 2(8 \div 2 \times (1+1))$	We have nested parentheses on the far right, so let's work on the inside set.
$= -2^2 + \sqrt{(6-2)} - 2(8 \div 2 \times 2)$	There aren't any more nested parentheses, so let's work on the set of parentheses on the far left.
$= -2^2 + \sqrt{4} - 2(8 \div 2 \times 2)$	Now, we'll work on the other set of parentheses. This set only has multiplication and division, so we'll work from left to right inside of the parentheses.
$= -2^2 + \sqrt{4} - 2(4 \times 2)$	Now, we'll complete that set of parentheses.
$= -2^2 + \sqrt{4} - 2(8)$	Let's rewrite slightly to completely get rid of all parentheses.
$= -2^2 + \sqrt{4} - 2 \times 8$	Now, we'll work on exponents, from left to right.
$= -4 + \sqrt{4} - 2 \times 8$	We only squared 2, and not -2. Square roots are really exponents, so we'll take care of that next.
$= -4 + 2 - 2 \times 8$	We're done with exponents; time for multiplication/division.
$= -4 + 2 - 16$	Now, only addition and subtraction are left, so we'll work from left to right.
$= -2 - 16$	Almost there!
$= -18$	

Our final answer is

$$-2^2 + \sqrt{6-2} - 2(8 \div 2 \times (1+1)) = -18$$

Computations with Rational Numbers

Rational numbers are real numbers that can be written as a fraction, such as $\frac{1}{2}$, $\frac{5}{4}$, and $-\frac{2}{3}$. You may notice that $\frac{5}{4}$ is a special type of rational number, called

---

Notes:

an *improper fraction*. It's called improper because the value in the numerator, 5, is bigger than the number in the denominator, 4. Often, students are taught to write these improper fractions as mixed numbers: $\frac{5}{4} = 1\frac{1}{4}$. This does help give a quick estimate of the value; we can quickly see that it is between 1 and 2. However, writing as a mixed number can make computations more difficult and can lead to some confusion when working with complicated expressions; it may be tempting to see $1\frac{1}{4}$ as $1 \times \frac{1}{4}$ rather than $\frac{5}{4}$. For this reason, we will leave all improper fractions as improper fractions.

With fractions, multiplication and exponents are two of the easier operations to work with, while addition and subtraction are more complicated. Let's start by looking at how to work with multiplication of fractions.

Example 10   **Multiplication of Fractions**
Evaluate

$$\frac{1}{4} \times \frac{2}{3} \times 2$$

Solution   With multiplication of fractions, we will work just like we do with any other type of real number and multiply from left to right. When multiplying two fractions together, we will multiply their numerators together (the tops) and we will multiply the denominators together (the bottoms).

$$\begin{aligned}
\frac{1}{4} \times \frac{2}{3} \times 2 &= \frac{1 \times 2}{4 \times 3} \times 2 \\
&= \frac{2}{12} \times 2 \\
&= \frac{1}{6} \times 2 \\
&= \frac{1}{6} \times \frac{2}{1} \\
&= \frac{1 \times 2}{6 \times 1} \\
&= \frac{2}{6} \\
&= \frac{1}{3}
\end{aligned}$$

After each step, we look to see if we can simplify any fractions. After the first multiplication, we get $\frac{2}{12}$. Both 2 and 12 have 2 as a factor (are both divisible by 2), so we can simplify by dividing each by 2, giving us $\frac{1}{6}$. Before doing the second multiplication, we transform the whole number, 2, into a fraction by writing it as $\frac{2}{1}$. Then, we can compute the multiplication of the two fractions, giving us $\frac{2}{6}$. We can simplify this because 2 and 3 have 2 as a common factor, so our final answer is

Notes:

$$\frac{1}{4} \times \frac{2}{3} \times 2 = \frac{1}{3}$$

Next, let's look at exponentiation of a fraction.

Example 11   **Exponentiation of a Fraction**

Evaluate

$$\left(\frac{1+2}{5}\right)^2$$

Solution   With exponentiation, we need to apply the exponent to both the numerator and the denominator. This gives

$$\left(\frac{1+2}{5}\right)^2 = \left(\frac{(1+2)}{(5)}\right)^2$$

$$= \left(\frac{(3)}{(5)}\right)^2$$

$$= \left(\frac{3}{5}\right)^2$$

$$= \frac{3^2}{5^2}$$

$$= \frac{9}{25}$$

Notice that we were careful to include the implied parentheses around the numerator and around the denominator. This helps to guarantee that we are correctly following the order of operations by working inside of any parentheses first, before applying the exponent. We can't simplify our fraction at any point since 3 and 5 do not share any factors. This gives us our final answer of

$$\left(\frac{1+2}{5}\right)^2 = \frac{9}{25}$$

Notes:

With division of fractions, we will build off of multiplication. For example, if we want to divide a number by 2, we know that we could instead multiply it by $\frac{1}{2}$ because dividing something into two equal pieces is the same as splitting it in half. These numbers are *reciprocals*; 2 can be written as $\frac{2}{1}$ and if we flip it, we get $\frac{1}{2}$, its reciprocal. This works for any fractions; if we want to divide by $\frac{5}{6}$, we can instead multiply by its reciprocal, $\frac{6}{5}$.

Addition and subtraction of fractions can be a bit more complicated. With a fraction, we can think of working with pieces of a whole. The denominator tells us how many pieces we split the item into, and the numerator tells us how many pieces we are using. For example, $\frac{3}{4}$ tells us that we split the item into 4 pieces and are using 3 of them. In order to add or subtract fractions, we need to work with pieces that are all the same size, so our first step will be getting a common denominator. We will do this by multiplying by 1 in a sneaky way. Multiplying by 1 doesn't change the meaning of our expression, but it will allow us to make sure all of our pieces are the same size.

Example 12   **Addition and Subtraction of Fractions**

Evaluate

$$\frac{1}{2} - \frac{1}{3} + \frac{1}{4}$$

Solution   Since we only have addition and subtraction, we will work from left to right. This means that our first step is to subtract $\frac{1}{3}$ from $\frac{1}{2}$. The denominators are different, so we don't yet have pieces that are all the same size. To make sure our pieces are all the same size, we will multiply each term by 1; we will multiply $\frac{1}{2}$ by $\frac{3}{3}$ and we will multiply $\frac{1}{3}$ by $\frac{2}{2}$. Since we are multiplying by the "missing" factor for each, both will have the same denominator. Once they have the same denominator, we can combine the numerators:

$$\frac{1}{2} - \frac{1}{3} + \frac{1}{4} = \frac{1}{2} \times \frac{3}{3} - \frac{1}{3} \times \frac{2}{2} + \frac{1}{4}$$

$$= \frac{3}{6} - \frac{2}{6} + \frac{1}{4}$$

$$= \frac{3-2}{6} + \frac{1}{4}$$

$$= \frac{1}{6} + \frac{1}{4}$$

$$= \frac{1}{6} \times \frac{2}{2} + \frac{1}{4} \times \frac{3}{3}$$

Notes:

$$= \frac{2}{12} + \frac{3}{12}$$

$$= \frac{2+3}{12}$$

$$= \frac{5}{12}$$

After combining the first two fractions, we had to find a common denominator for the remaining two fractions. Here, we found the smallest possible common denominator. We did this by looking at each denominator and factoring them. The first denominator, 6, has 2 and 3 as factors; the second denominator, 4 has 2 as a repeated factor since $4 = 2 \times 2$. These means our common denominator needed to have 3 as a factor and 2 as a double factor: $3 \times 2 \times 2 = 12$. We don't have to find the smallest common denominator, but it often keeps the numbers more manageable. We could have instead done:

$$\frac{1}{6} + \frac{1}{4} = \frac{1}{6} \times \frac{4}{4} + \frac{1}{4} \times \frac{6}{6}$$

$$= \frac{4}{24} + \frac{6}{24}$$

$$= \frac{4+6}{24}$$

$$= \frac{10}{24}$$

$$= \frac{5}{12}$$

We still end up with the same final answer:

$$\boxed{\frac{1}{2} - \frac{1}{3} + \frac{1}{4} = \frac{5}{12}}$$

Like fractions, decimals can also be difficult to work with. Note that all repeating decimals and all terminating decimals can be written as fractions: $0.\overline{333} = \frac{1}{3}$ and $2.1 = 2 + \frac{1}{10} = \frac{20}{10} + \frac{1}{10} = \frac{21}{10}$. You can convert these into fractions or you can work with them as decimals. When adding or subtracting decimals, make sure to align the numbers at the decimal point. When multiplying, first multiply as though there are no decimals, aligning the last digit of each number. Then, as your final step place the decimal point so that it has the appropriate number

Notes:

of digits after it. For example,

$$
\begin{array}{r}
1.2 \\
\times \quad 1.1\,5 \\
\hline
6\,0 \\
1\,2 \\
1\,2 \\
\hline
1.3\,8\,0
\end{array}
$$

Because 1.2 has one digit after the decimal place and 1.15 has 2 digits after the decimal place, we need a total of $1+2=3$ digits after the decimal place in our final answer, giving us 1.380, or 1.38. It is important to note that we placed the decimal point before dropping the zero on the end; our final answer would have quite a different meaning otherwise.

Computations with Units

So far, we have only looked at examples without any context to them. However, in calculus you will see many problems that are based on a real world problem. These types of problems will come with units, whether the problem focuses on lengths, time, volume, or area. With these problems, it is important to include units as part of your answer. When working with units, you first need to make sure all units are consistent; for example, if you are finding the area of a square and one side is measured in feet and the other side in inches, you will need to convert so that both sides have the same units. You could use measurements that are both in feet or both in inches, either will give you a meaningful answer. Let's look at a few examples.

Example 13   Determining Volume
Determine the volume of a rectangular solid that has a width of 8 inches, a height of 3 inches, and a length of 0.5 feet.

Solution   First, we need to get all of our measurements in the same units. Since two of the dimensions are given in inches, we will start by converting the third dimension into inches as well. Since there are 12 inches in a foot, we get

$$
0.5\text{ ft} \times \frac{12\text{ in}}{1\text{ft}} = \frac{0.5 \times 12\text{ ft} \times \text{ in}}{1\text{ ft}}
$$
$$
= \frac{6\text{ ft} \times \text{ in}}{1\text{ ft}}
$$
$$
= 6\text{ in}
$$

In the last step we simplify our fraction. We can simplify $\frac{6}{1}$ as 6, and we can

Notes:

simplify $\frac{ft \times in}{ft}$ as in. This means that we know our rectangular solid is 8 inches wide, 3 inches tall, and 6 inches long. The volume is then

$$
\begin{aligned}
V &= (8 \text{ in}) \times (3 \text{ in}) \times (6 \text{ in}) \\
&= (8 \times 3 \times 6) \times (\text{ in} \times \text{ in} \times \text{ in}) \\
&= (24 \times 6) \times (\text{ in} \times \text{ in} \times \text{ in}) \\
&= 144 \text{ in}^3
\end{aligned}
$$

Since all three measurements are in inches and are being multiplied, we end up with units of inches cubed, giving us a final answer of

$$
\boxed{V = 144 \text{ in}^3}
$$

Units can also give you hints as to how a number is calculated. For instance, the speed of a car is often measured in mph, or miles per hour. We write these units in fraction form as $\frac{miles}{hour}$, which tells us that in our computations we should be dividing a distance by a time. Sometimes, however, a problem will start with units, but the final answer will have no units, meaning it is unitless. We will run across examples of this when we discuss trigonometric functions. Trigonometric functions can be calculated as a ratio of side lengths of a right triangle. For example, in a right triangle with a leg of length 3 inches and a hypotenuse of 5 inches, the ratio of the leg length to the hypotenuse length is $\frac{3 \text{ in}}{5 \text{ in}} = \frac{3}{5}$. Since both sides are measured in inches, the units cancel when we calculate the ratio. We would see the same final answer if the triangle had a leg of 3 miles and a hypotenuse of 4 miles; they are similar triangles, so the ratios are the same.

In this section, we have examined how to work with basic mathematical operations and how these operations interact with each other. In the next section we'll talk about how to make specialized rules or operations through the use of functions. In the exercises following this section, we continue our work with order of operations and practice these rules in situations with a bit more context. Note that answers to all example problems are available at the end of this book to help you gauge your level of understanding. If your professor allows it, it is a good idea to check the answer to each question as you complete it; this will allow you to see if you understand the ideas and will prevent you from practicing these rules incorrectly.

Notes:

Exercises 1.1

Terms and Concepts

1. In your own words, what does "multiplication and division are on the same level" mean?

2. In an expression with both addition and multiplication, which operation do you complete first?

3. In your own words, what is meant by "implied parentheses"? Provide an example.

4. T/F: In an expression with only addition and subtraction remaining, you must complete all of the addition before starting the subtraction. Explain.

5. T/F: In the expression -2^2, only "2" is squared, not "-2." Explain.

Problems

In exercises 6 – 15, simplify the given expressions.

6. $-2(11 - 5) \div 3 + 2^3$

7. $\dfrac{3}{5} + \dfrac{2}{3} \div \dfrac{5}{6}$

8. $\left(\dfrac{2}{3}\right)^2 - \dfrac{1}{6}$

9. $(13 + 7) \div 4 - 3^2$

10. $\left(5 - \dfrac{1}{2}\right)^3$

11. $(2)(-2) \div \dfrac{1}{2}$

12. $\dfrac{2 - 4(3 - 5)}{6 - 7 + 3} - \sqrt{25 - 9}$

13. $\sqrt{\dfrac{2^2 + 3^2 + 5}{2 - (-1)^3}} - 2 + 6$

14. $\dfrac{4 - 1(-2)}{\frac{1}{2} + 1} - 2$

15. $-4^2 + 5^2 - 2 \times 3 + 1$

In exercises 16 – 21, evaluate the described mathematical statement, or determine how the described changes affect other variables in the statement as appropriate.

16. A runner leaves her home and runs straight for 30 minutes at a pace of 1 mi every 10 minutes (a 10-minute mile). She makes a 90-degree left turn and then runs straight for 40 minutes at the same pace. What is the distance between her current location and her home?

17. The Reynold's number, which helps identify whether or not a fluid flow is turbulent, is given by $Re = \frac{\rho u D}{\mu}$. If ρ, u, and D are held constant while μ increases, does Re increase, decrease, or stay the same?

18. Consider a square-based box whose height is twice the length of the base of the side. If the length of the square base is 3 ft, what is the volume of the box? (Don't forget your units!)

19. The velocity of periodic waves, v, is given by $v = \lambda f$ where λ is the length of the waves and f is the frequency of the waves. If the wavelength is held constant while the frequency is tripled, what happens to the velocity of the waves? Be as descriptive as possible.

20. The capacitance, C, of a parallel plate capacitor is given by $C = \frac{k\varepsilon_0 A}{d}$ where d is the distance between the plates. If k, ε_0, and A are held constant while the distance between the plates decreases, does the capacitance increase, decrease, or stay the same?

21. Consider a square-based box whose height is half the length of the sides for the base. If the surface area of the base is 16 ft^2, what is the volume of the box?

In exercises 22 – 25, evaluate the given work for correctness. Is the final answer correct? If not, describe the error(s) in the solution.

22.

$$
\begin{aligned}
12 \div 6 \times 4 - (3 - 4)^2 &= 12 \div 6 \times 4 - (-1)^2 \\
&= 12 \div 6 \times 4 + 1 \\
&= 12 \div 24 + 1 \\
&= \frac{1}{2} + 1 \\
&= \frac{3}{2}
\end{aligned}
$$

23.

$$
\begin{aligned}
-3^2 + 6 \div 2 + (-4)^2 &= -9 + 6 \div 2 + (-4)^2 \\
&= -9 + 6 \div 2 + 16 \\
&= -9 + 3 + 16 \\
&= 10
\end{aligned}
$$

24.

$$\frac{2+(3-5)}{2-2(6+3)} + \sqrt{64+36} = \frac{2+(-2)}{2-2(6+3)} + \sqrt{64+36}$$

$$= \frac{2-2}{2-2(9)} + \sqrt{64+36}$$

$$= \frac{2-2}{2-2(9)} + 8 + 6$$

$$= \frac{2-2}{2-18} + 8 + 6$$

$$= \frac{0}{-16} + 14$$

$$= 14$$

25.

$$\frac{(-3+1)(2(-3)) - ((-3)^2-3)(1)}{(-3+1)^2} = \frac{(-3+1)(2(-3)) - ((-3)^2-3)(1)}{(-2)^2}$$

$$= \frac{(-2)(-6) - ((-3)^2-3)(1)}{(-2)^2}$$

$$= \frac{(-2)(-6) - (9-3)(1)}{(-2)^2}$$

$$= \frac{(-2)(-6) - (6)(1)}{4}$$

$$= \frac{-12-6}{4}$$

$$= \frac{-18}{4}$$

$$= \frac{-9}{2}$$

1.2 Introduction to Functions

This section introduces ideas and notation for functions. Much of the work in calculus relies heavily on understanding the meaning of a function and a proper understanding of function notation. Here we'll talk about these ideas and work through several examples involving function notation and how they relate to calculus.

What is a Function?

In mathematics, we look for patterns to help explain the world around us. Mathematicians often use functions to express these patterns succinctly. For example, we learn in geometry that the area of a square with sides of length 2 in is $2 \times 2 = 4$ in^2. Similarly, if the square has sides of length 3 in, it's area is $3 \times 3 = 9$ in^2. This shows us a pattern for determining the area of a square: if we know the side length, we simply multiply the side length by itself to get the area. Rather than writing out what this rule looks like for all sorts of different side lengths, we can express the pattern as a function:

$$\begin{aligned} A(x) &= x \times x \\ &= x^2 \end{aligned} \tag{1.13}$$

This function tells us that the area of a square with sides of length x has an area of x^2. This is a lot more compact than writing out a table with all sorts of different side lengths and areas.

Function Notation

Here we say that x is the *input* of the function $A(x)$ (read as "A of x"), and that x^2 is the corresponding *output*. Notice that since we get to choose the "name" of the function, A, we used something that has some meaning for our example; our function gives us area, so calling the function A makes that clearer than if we had chosen something like $l(x)$, where you might be tempted to think l for length.

Mathematicians will often use letters like f, g, and h to name their functions, but you can name your functions anyway you like. In fact, some functions that you may already be familiar with have longer names, like sin for the sine function or cos for the cosine function. Similarly, mathematicians will often use x to represent the input of the function, but you can choose any name you want. In our example above, we could use l as our input to stand for "length", giving us $A(l) = l^2$. This looks a little different than using $A(x)$, but it provides the same meaning for mathematicians: take your input and square it

Notes:

Often, mathematicians will use "the function A" and "the function $A(x)$" interchangeably. Both tell us to use the same rule that is shown in (1.13), but the second gives us an added bit of information; it tells us that for the function A, x is our input variable. For our example function, this information isn't particularly useful because the only letter on the right side of our function is x, but some functions will have other letters that aren't input variables. We'll run into this fairly often in calculus. For example, suppose we want to know the height of a ball that has been thrown into the air. Physics (and calculus) gives us a function for this:

$$h(t) = h_0 + v_0 t + \frac{1}{2}at^2 \qquad (1.14)$$

Since the left side has $h(t)$, we know that t is our input variable, but we have lots of other letters on the right side. These letters all have meaning for this problem: h_0 is the initial height of the ball, v_0 is the velocity it was thrown at, and a is the acceleration due to gravity. While they all have meaning and can change based on the particular instance of a ball being thrown, they are considered *parameters* of the function, and not input variables. Why? Well, as soon as the ball is thrown, h_0, v_0, and a won't change for that ball. Only the time the ball has been in the air changes; the ball has a different height after $t = 2000$ seconds than it did after only $t = 2$ seconds. Therefore, only t is an input variable for this function. While it might not seem like a big deal to write A instead of $A(x)$, we can see that writing h instead of $h(t)$ could lead to confusion, so it's good to be careful and include that input variable when it's not perfectly clear.

Evaluating a Function

Now that we are familiar with why we use functions let's look at how to evaluate a function. We'll start with evaluating a function for a single value.

Example 14   **Evaluating a Function at a Point**
Determine the value of $f(-2)$ if $f(x) = x^2 + 4x - 10$.

Solution   First, we notice that the left side tells us that our input is x. Since we want to determine the value of $f(-2)$, we'll replace every x on the right side with (-2).

$$\begin{aligned} f(-2) &= (-2)^2 + 4(-2) - 10 \\ &= (4) + 4(-2) - 10 \\ &= 4 - 8 - 10 = -14 \end{aligned} \qquad (1.15)$$

So, we find that

Notes:

$$f(-2) = -14$$

It's good to notice that the question in Example 14 can be written in several different ways. All of the following require the same work, but are worded in slightly different ways:

- Determine the value of $f(-2)$

- Determine the value of $f(x)$ for $x = -2$

- Evaluate $f(-2)$

- Evaluate f at -2

There are probably more ways to ask this question, but these are some of the most common ones. Let's look at an example where the function has parameters.

Example 15   Evaluating a Function with Parameters
Using the height formula in equation (1.14), determine the height of a ball 5 seconds after it was thrown.

Solution   First, let's make sure we have the correct equation. The (1.14) label is next to $h(t) = h_0 + v_0 t + \frac{1}{2}at^2$, so that tells us we are working with that function. The left side tells us that t is our input variable since the function is called $h(t)$. That means we need to substitute (5) for t everywhere t appears in the function:

$$h(5) = h_0 + v_0(5) + \frac{1}{2}a(5)^2$$
$$= h_0 + 5v_0 + \frac{1}{2}a(25)$$
$$= h_0 + 5v_0 + \frac{25}{2}a$$

Notice that our answer includes all three parameters. This is to be expected because we weren't given values for these parameters, so we'll leave them as letters rather than making up numbers to use. This gives us the flexibility to determine the height after 5 seconds for a variety of parameter values, and gives a final answer of

---

Notes:

$$h(5) = h_0 + 5v_0 + \frac{25}{2}a$$

Notice that in Example 14, we replaced x not just with -2, but with (-2) and in Example 15 we replaced t with (5). This helps in a couple of ways. First, it makes sure we don't miss any implied parentheses when we square x in Example 14. Second, it makes sure we replace x and t with the *entire* input. This becomes very important in calculus. In differential calculus, you will spend a lot of time looking at how quickly function outputs change when the input only changes a tiny bit. You will do this by looking at a *difference quotient* for the function. The general form of difference quotient for the function $f(x)$ that you will use is:

$$\frac{f(x + h) - f(x)}{h} \tag{1.16}$$

Notice that the numerator starts with $f(x + h)$. This means that every x on the right side needs to be replaces with $x + h$. Here, the parentheses make a big difference even with a simple function like $p(x) = x^2$. If we include the parentheses, we get that

$$\begin{aligned} p(x + h) &= (x + h)^2 \\ &= (x + h) \times (x + h) \\ &= x^2 + 2xh + h^2 \end{aligned} \tag{1.17}$$

However, if we don't include the parentheses, we would get $x + h^2$, which is a very different (and incorrect) answer. Let's look at an example of finding a difference quotient for a more complicated function.

Example 16 **Finding a Difference Quotient**
Find the difference quotient for $g(t) = 2t^2 - 3t + 1$

   Solution   Here we have a function called g, with t as its input. That means that in our difference quotient, we will have g instead of f and t instead

Notes:

of x, but h will still be h. So, our difference quotient will look like

$$
\begin{aligned}
\frac{g(t+h) - g(t)}{h} &= \frac{\left[2(t+h)^2 - 3(t+h) + 1\right] - \left[2t^2 - 3t + 1\right]}{h} \\
&= \frac{\left[2(t^2 + 2th + h^2) - 3(t+h) + 1\right] - \left[2t^2 - 3t + 1\right]}{h} \\
&= \frac{\left[2t^2 + 4th + 2h^2 - 3t - 3h + 1\right] - \left[2t^2 - 3t + 1\right]}{h} \\
&= \frac{2t^2 + 4th + 2h^2 - 3t - 3h + 1 - \left[2t^2 - 3t + 1\right]}{h} \\
&= \frac{2t^2 + 4th + 2h^2 - 3t - 3h + 1 - 2t^2 + 3t - 1}{h} \\
&= \frac{4th + 2h^2 - 3h}{h} \\
&= 4t + 2h - 3
\end{aligned}
$$

There are a few important things to notice here. First, when we replaced t with $(t + h)$ in the first term, we included those parentheses to make sure we used the whole input. Second, from line 3 to line 5, we dropped all parentheses; when we did this we made sure to distribute the negative to everything inside the second set of parentheses, and not just the first term. We end up with a final answer of

$$
\boxed{\frac{g(t+h) - g(t)}{h} = 4t + 2h - 3}
$$

Common Types of Functions

There are several different types of functions that get use commonly in calculus. In this section, we'll briefly describe each. Later, we'll talk about how we can combine these in different ways, what types of inputs these functions can take, and what their graphs look like.

Power Functions

A *power function* is any function that involves a variable raised to a power:

$$
f(x) = ax^b \tag{1.18}
$$

Notes:

Here, the left side tells us that x is the variable; a and b are parameters that can be any real numbers. Because a and b can be anything, this is a very general function type meaning that the properties of the function can be very different based on these values of a and b.

A *monomial* is a special type of power function where b is a non-negative integer; this means b can be 0, 1, 2, 3, etc. We call b the *degree* of the function. $f(x) = 2x$ has degree 1, $g(x) = 45x^{13}$ has degree 13, and $h(x) = 12 = 12x^0$ has degree 0. Later, we'll see that the degree helps us to quickly determine the shape of the function when we graph it.

If we take one or more monomials and add them together, we get a *polynomial*. That means that $f(x) = 2x$, $g(x) = 45x^{13}$, and $h(x) = 12 = 12x^0$ are not only monomials, but also polynomials, and if we add them all together we get a new polynomial: $p(x) = 45x^{13} + 2x + 12$. We could get a different polynomial by taking the difference (subtracting) them: $q(x) = -45x^{13} - 2x - 12$. There are many more polynomials we could make from the three functions with various combinations of addition and subtraction.

Traditionally, polynomials are written with the highest degree monomial first because for big values of x it becomes the most important term. The highest degree monomial also tells us the degree of the polynomial: $p(x)$ and $q(x)$ both have degree 13. If the degree of the polynomial is 3, like with $r(\theta) = 4\theta^3 + 2\theta^2 - 5\theta + 2$, we can call it a *cubic* function, and if the degree is 2, we call it a *quadratic* function. If the degree is 1, like with $n(t) = 5t - 2$, we simply call it a *linear* function, and if the degree is 0, we say it's a *constant* function. These four all have special names because they get used very often in mathematics.

Root Functions

Later, we'll talk more about the importance of root functions, but for now we'll focus on what they look line in their general form. A *root* function is any function that looks like $f(x) = x^{1/n}$ where n is a natural number (a positive integer, or counting number like 1, 2, 3, 4, etc.). This means that root functions are a special type of power function. The most commonly used root function is the square root function, $f(x) = x^{1/2}$. You've most likely seen this written in a different form: $f(x) = \sqrt{x}$. There are many other root functions like the cube root function ($g(x) = x^{1/3} = \sqrt[3]{x}$) and the fourth root function ($h(x) = x^{1/4} = \sqrt[4]{x}$). In general, we say that $x^{1/n}$ is the nth root of x, so $x^{1/7}$ would be called the seventh root of x. These can sometimes be tricky to evaluate. You probably know that $9^{1/2} = \sqrt{9} = 3$ because $3^2 = 9$, but few people know a good approximation for $5^{1/2}$. In these situations, it's usually best to leave your answer as $5^{1/2}$ or $\sqrt{5}$ rather than using a calculator to turn it into a decimal because it's more precise (and quicker to write than 2.2360679775).

Notes:

Exponential Functions

Exponential functions have the form $f(x) = b^x$, with $b > 0$ and $b \neq 1$. Notice that like a power function, an exponential function involves an exponent, but there is a big difference. For a power function, the input variable, x is the base with a parameter as the exponent. For an exponential function, the roles are swapped: the base is a parameter and the input variable is the exponent.

Logarithmic Functions

Logarithms, or *logarithmic* functions are quite important in many applications of calculus because each logarithmic function is the inverse of an exponential function. They have the form $f(x) = \log_b(x)$. Just like with exponentials, we need $b > 0$ and $b \neq 1$. There are two very commonly used logarithms. The first is $\log_{10}(10)$, read as "log base 10 of x." Sometimes you will see this written as just $\log(x)$ instead of $\log_{10}(x)$. The second commonly used logarithm is $\log_e(x)$, "log base e of x", also know as the natural logarithm (commonly written as $\ln(x)$).

Trigonometric Functions

Trigonometric functions are functions that relate the angles of a triangle to the length of the sides in that triangle. They can also be used to describe many natural phenomena like waves (sound, light, and water waves) and harmonic motion (motion that repeats the same pattern over and over, also know as cyclic motion). The trigonometric functions that are most commonly used are sine ($\sin(x)$), cosine ($\cos(x)$), and tangent ($\tan(x)$). We'll talk about these functions and their application more later in this text.

Combining Functions

While each of these function types has its own set of special uses, often combinations of these functions are needed to accurately model events. For this section, we will use three different functions to help provide examples of how we can combine and modify functions:

$$f(x) = 3x^2 \tag{1.19}$$

$$g(x) = x - 4 \tag{1.20}$$

$$h(x) = \sqrt{x} + 6 \tag{1.21}$$

In differential calculus it is very important to be able to recognize how functions are combined. How they are combined greatly impacts how you take the

Notes:

derivative of the function. This text will not cover derivatives, but they are one of the most important topics in calculus, so being able to recognize these combination methods will be quite useful in calculus.

Scalar Multiples of Functions

The first way we can modify functions is with *scalar multiplication*. This simply means multiplying the function by a constant (a number). For example,

- $4f(x) = 4(3x^2) = 12x^2$;

- $4g(x) = 4(x - 4) = 4x - 16$;

- $4h(x) = 4(\sqrt{x} + 6) = 4\sqrt{x} + 24$.

There's nothing special about the number 4, we could multiply by anything: negative numbers, positive numbers, whole numbers, fractions, decimals, or even zero (even though that would make for a pretty boring result). Notice that for each of these we used parenthesis around the whole function when we multiplied. This makes sure that we really multiplied the entire function by 4, and not just part of the function. This is particularly important with $g(x)$ and $h(x)$ since they each had two terms already and we had to distribute the 4 to both terms.

Sums and Differences of Functions

One way to combine functions is to add them (sums) or subtract them (differences) from each other. For example, the sum of $f(x)$ and $g(x)$ is $f(x) + g(x) = 3x^2 + x - 4$. Sums are nice to work with for many reasons; mathematicians use sums of functions to get better and better approximations when working with complicated data, and with sums order doesn't change the result. If we did $g(x) + f(x)$ instead of $f(x) + g(x)$, we get $g(x) + f(x) = x - 4 + 3x^2$; if we rearrange terms so that the highest degree comes first, we get $3x^2 + x - 4$ which is exactly the same as $f(x) + g(x)$.

With differences, we have to be a little more careful because the order will make a difference. Let's take a look:

- $f(x) - g(x) = 3x^2 - (x - 4) = 3x^2 - x + 4$

- $g(x) - f(x) = x - 4 - (3x^2) = x - 4 - 3x^2 = -3x^2 + x - 4$

Here we see that $f(x) - g(x)$ and $g(x) - f(x)$ give us different results. Like with scalar multiplication, we were again careful to put parentheses around the entire function when we wrote the second function. This is because subtracting

Notes:

it really involves multiplying it by -1 and we want to make sure we distribute that negative to the entire function.

With both sums and differences, we can use as many functions as we want:

$$f(x)-h(x)-g(x) = 3x^2-(x-4)-(\sqrt{x}+6) = 3x^2-x+4-\sqrt{x}-6 = 3x^2-x-\sqrt{x}-2$$

We can also mix between addition and subtraction:

$$h(x) + g(x) - f(x) = \sqrt{x} + 6 + x - 4 - (3x^2) = -3x^2 + x + \sqrt{x} + 2$$

Products of Functions

Another way of combining functions is through products (multiplication) of functions. Like with sums of functions, order doesn't make a difference, so $f(x)g(x) = g(x)f(x)$. We won't show the details here, but try to verify it on your own. (Note: it is common for college level mathematics textbooks to state a property like this without showing the details. This means that the author(s) believe you are capable of working through the steps on your own, and working through these statements is a good way to verify that you do understand the steps involved.) With products of functions, we again will want to use parentheses to make sure we are using the entire function as one unit. This is particularly important when the function has multiple terms:

- $g(x)f(x) = (x - 4)(3x^2) = (x)(3x^2) - 4(3x^2) = 3x^3 - 12x^2$
- $h(x)g(x) = (\sqrt{x}+6)(x-4) = (\sqrt{x})(x-4)+6(x-4) = x\sqrt{x}-4\sqrt{x}+6x-24$

Quotients of Functions

Next, we can combine functions by through division. We call the function $\frac{f(x)}{g(x)}$ the *quotient* of f and g. As with differences, order matters here; the quotient of f and g is different than the quotient of g and f. (Reminder: this is another good place to try verifying a property on your own. Showing that things are different can be just as useful as showing that they are the same.) Remember that with fractions we have implied parenthesis around the entire numerator and around the entire denominator so we don't need to explicitly include those parentheses here. Typically we won't have to worry about much simplification with quotients of functions; later we'll see how to identify when we may be able to simplify, but for now it's safer to *not* simplify these types of combinations. Let's look at a few examples:

- $\dfrac{f(x)}{g(x)} = \dfrac{3x^2}{x - 4}$

Notes:

- $\dfrac{g(x)}{f(x)} = \dfrac{x-4}{3x^2}$

- $\dfrac{h(x)}{g(x)} = \dfrac{\sqrt{x}+6}{x-4}$

Composition of Functions

The last way we can combine functions is quite different. With all of our previous methods, we could take the output from one function and use arithmetic to combine it with the output from another function. For example, if we wanted to know $f(4) + g(4)$ but didn't care about the function $f(x) + g(x)$ in general, we could simply find $f(4)$ ($f(4) = 3(4^2) = 3(16) = 48$) and $g(4)$ ($g(4) = (4) - 4 = 0$) and add them together: $f(4) + g(4) = 48 + 0 = 48$. With composition of functions, we are going to use the output of one function as the input for another function. The *composition* of $f(x)$ with $g(x)$ is written as $f(g(x))$, or as $(f \circ g)(x)$, using mathematical notation and is read as " f of g of x." If we look at the notation, we see that function f is going to take $g(x)$ as it's input variable. $g(x)$ will sometimes be referred to as the "inside" function and $f(x)$ as the "outside" function because $g(x)$ goes "inside" of f. As an example, let's look at $f(g(4))$ ("f of g of 4"). This tells us that we want to find the value of f when we input $g(4)$. Well, we know from above that the value of $g(4)$ is 0, so let's see what happens when we input 0 into f. We would get that $f(0) = 3(0^2) = 3(0) = 0$. To show this work using only mathematical notation, we would write

$$
\begin{aligned}
f(g(4)) &= f(0), \text{ since } g(4) = 0 \\
&= 3(0^2) \\
&= 3(0) \\
&= 0
\end{aligned}
\tag{1.22}
$$

That's great if we just care about one point, but what if we want to know what $f(g(x))$ looks like at several different points?

Example 17   **Composing Two Functions**
Using $f(x)$ and $g(x)$ from above, determine $j(x) = f(g(x))$.

   Solution   Since $g(x)$ is our input, we need to replace every x in f with

---

Notes:

$(x - 4)$. This gives us

$$
\begin{aligned}
j(x) = f(g(x)) &= f(x - 4) \\
&= 3(x - 4)^2 \\
&= 3(x - 4)(x - 4) \\
&= 3(x^2 - 8x + 16) \\
&= 3x^2 - 24x + 48
\end{aligned}
\tag{1.23}
$$

Our final result is

$$
\boxed{j(x) = 3x^2 - 24x + 48}
$$

We can verify that this agrees with the single point we looked earlier:

$$
\begin{aligned}
j(4) &= 3(4^2) - 24(4) + 48 \\
&= 3(16) - 24(4) + 48 \\
&= 48 - 96 + 48 \\
&= 0
\end{aligned}
$$

Composition of functions is another place where order can make a difference. Let's take a look at $g(f(x))$.

Example 18   **Composing Two Functions**

Using $f(x)$ and $g(x)$ from above, determine $k(x) = g(f(x))$.

Solution   Since $f(x)$ is our input, we need to replace every x in g with $(3x^2)$. This gives us

$$
\begin{aligned}
k(x) = g(f(x)) &= g(3x^2) \\
&= (3x^2) - 4 \\
&= 3x^2 - 4
\end{aligned}
\tag{1.24}
$$

Our final result is

$$
\boxed{k(x) = 3x^2 - 4}
$$

Notes:

We can see that $j(x)$ and $k(x)$ are very different functions; we already saw that $j(4) = 0$, and we can see that $k(4) = 3(4)^2 - 4 = 3(16) - 4 = 48 - 4 = 44$.

Function composition is not limited to using different functions for the inside function and the outside function. We could look at compositions like $f(f(x))$, $g(g(x))$, or $h(h(x))$. We work with these the same way we worked with $f(g(x))$ and $g(f(x))$; replace every x in the outside function with the entire inside function. Nor is function composition restricted to only two functions; we could look at compositions with many layers. Let's take a look at an example with 3 layers.

Example 19   **Composing Three Functions**
Using $f(x)$ and $g(x)$ from above, determine $m(x) = f(g(g(x)))$.

Solution   With multiple layers of composition, it's typically easiest to start on the inner layer first and then work your way out. Here the outermost function is $f(x)$, then $g(x)$ in the middle, and $g(x)$ on the inside. We already know what $g(x)$ looks like by itself, and the first composition we run into is $g(g(x))$. Let's call this $m_{inside}(x)$:

$$\begin{aligned} m_{inside}(x) = g(g(x)) &= g(x - 4) \\ &= (x - 4) - 4 \\ &= x - 4 - 4 \\ &= x - 8 \end{aligned} \qquad (1.25)$$

Just like before, we took the inside function, $(x - 4)$ and used it to replace every x in the outside function. Now, we've done the first layer of composition. We can now write $m(x) = f(g(g(x)) = f(m_{inside}(x))$. Now we have one last composition to worry about, with m_{inside} as the inside function and f as the outside function:

$$\begin{aligned} m(x) = f(m_{inside}(x)) &= f(x - 8) \\ &= 3(x - 8)^2 \\ &= 3(x - 8)(x - 8) \\ &= 3(x^2 - 16x + 64) \\ &= 3x^2 - 48x + 192 \end{aligned} \qquad (1.26)$$

This gives our final result:

$$\boxed{m(x) = 3x^2 - 48x + 192}$$

Notes:

Multiple Combinations of Functions

We've talked about many different ways to combine functions. It is important to note that all of the combination methods can be mixed together. We could create a combination like $f(x)[g(x)+h(x)]$ where we add g and h and then multiply the result with f, or a combination like $g(2f(x))$ where we multiply f by a scalar and then use that as the input for g. As when we work with numbers, we must still use our same order of operations rules when we work with functions. For example, in the combination $[f(x) + g(x)][h(g(x))]$ we would need to complete the combinations inside of each set of brackets before multiplying the results.

Notes:

Exercises 1.2

Terms and Concepts

1. What does "the function $f(t)$" tell you that "the function f" does not?

2. T/F: If $g(x) = x^2$, then $g(2) = g(-2)$.

3. T/F: You can't combine functions using both composition and quotients in the same function.

4. T/F: In the combination $g(f(x))$, $f(x)$ is the input for $g(x)$.

Problems

Let $f(x) = x^3$, $g(x) = x + 4$, and $h(x) = \sin(x)$. Each of exercise 5 – 8 is some combination of $f(x)$, $g(x)$, and $h(x)$. Determine the type of combination and write it using function notation. For example, $x^3 + x + 4$ is the addition of $f(x)$ and $g(x)$ and can be written as $f(x) + g(x)$.

5. $\frac{x^3}{\sin(x)}$

6. $\sin(x + 4)$

7. $\sin(x) + 4$

8. $(2x^3)(x + 4)$

In exercises 9 – 11, determine the input variable of each function, any parameters of the function, and the type of function.

9. $C(A) = \frac{k\varepsilon_0 A}{d}$

10. $v(t) = -9.8t + v_0$

11. $A(t) = P(1 + \frac{r}{n})^{nt}$

In exercises 12 – 17, evaluate the given expression.

12. Given $f(x) = 2x^2$ and $g(x) = x - b$, find $5f(3a) - g(4)$

13. Given $f(x) = x^2 - 3$ and $g(x) = x - b$, find $f(y + h) - 3g(5)$

14. Given $f(x) = 5 - x$ and $g(x) = -x^4 + p$, find $f(y + h) - 3g(y)$

15. Given $f(\theta) = \frac{\theta + 3}{\theta - 2}$ and $g(\theta) = \theta^2 + 4$, find $g(f(3))$

16. Given $g(x) = x^2 - 4$ and $f(x) = \sqrt{x + 8}$, find $g(x + h) - 2f(8)$

17. Given $f(y) = y - 5$ and $g(y) = h - y^2$, find $g(f(y)) - f(g(y))$

In exercises 18 – 21, determine the difference quotient of each of the following functions.

18. $h(r) = 2r + 4$

19. $g(y) = 4y - 7$

20. $y(x) = x^2 + 6$

21. $f(t) = 4t^2 + x$

1.3   Factoring and Expanding

First, we will look at how to correctly expand a product of polynomials. Once we have discussed this skill, we will look at factoring polynomials. Expanding and factoring are inverse ideas; both work with the same two forms and help us switch back and forth between these two forms. Expanding works off of the ideas we saw when we looked at order of operations, but typically involves variables or parameters in such a way that we can't write the expression without using addition or subtraction.

Expanding

When we learn how to multiply two two-digit numbers together, we are using the same ideas that get used in expanding. Let's take our first look at how we will expand products of functions by seeing those methods, but with multiplying two two-digit numbers together instead of multiplying two functions. This will show you the methods we will use, but with a problem you already know how to do. These methods will show you a new way of looking at this problem that will help us expand functions correctly.

Example 20   **Multiplying Two Two-Digit Numbers**
Evaluate $(40 + 2)(30 + 1)$.

Solution   Typically, we would start this problem by looking at our order of operations. Our order of operations tells us to do everything inside the parentheses first, which would give us $(42)(31)$, and then we would multiply these. However, we are going to use the *distributive property* instead. The distributive property tells us that every term in the first set of parentheses must get multiplied with the second set of parentheses:

$$
\begin{aligned}
(40 + 2)(30 + 1) &= 40(30 + 1) + 2(30 + 1) \\
&= 40 \times 30 + 40 \times 1 + 2 \times 30 + 2 \times 1 \\
&= 1200 + 40 + 60 + 2 \\
&= 1302
\end{aligned}
\tag{1.27}
$$

After multiplying the second set of parentheses by every term in the first set, we then use the distributive property again. In this second step, we distributed 40 to both 30 and 1 and distributed 2 to both 30 and 1. After these multiplications are formed, we end up with four terms. Here, all four terms are just numbers and can be added together to get the final answer:

Notes:

$$(40 + 2)(30 + 1) = 1302$$

Clearly, for this problem, this is not the easiest way to get the final answer, but it illustrates how we can correctly use the distributive property. Use of the distributive property becomes very important when we have variables or parameters involved and can't simplify inside of the parentheses.

Example 21   **Expanding the Product of Linear Functions**
Expand $f(x)g(x)$, where $f(x) = 2x - 1$ and $g(x) = x + 5$.

   Solution   First, we need to make sure we are correctly using parentheses in this problem. We want to expand the product of $f(x)$ and $g(x)$, each of which has two terms. This means that we need to include a set of parentheses around $f(x)$ and a set around $g(x)$ to make sure the we multiply with the whole function. After that, we will use the distributive property, just like we did in the previous example.

$$\begin{aligned} f(x)g(x) &= (2x - 1)(x + 5) \\ &= 2x(x + 5) - 1(x + 5) \\ &= 2x^2 + 10x - x - 5 \\ &= 2x^2 + 9x - 5 \end{aligned} \qquad (1.28)$$

Just like in our previous example, we distributed by first multiplying each term from the first set of parentheses to the second set of parentheses. In the end, we were able to combine like-terms because we had two linear terms: $10x$ and $-x$. No other terms could be combined because there was only one quadratic term and only one constant term, giving us a final answer of

$$f(x)g(x) = 2x^2 + 9x - 5$$

   Many people will skip the step of writing out $2x(x + 5) - 1(x + 5)$ and will jump directly to $2x^2 + 10x - x - 5$. One way you can make this jump is by using the acronym FOIL. FOIL stands for First, Outside, Inside, Last. It says that you should multiply the first term from each set of parentheses together, then the "outside" terms, then the "inside" terms, and then the last terms. This works very well when each set of parentheses only has two terms in it. However, if the parentheses have more than two terms each, FOIL can be a bit misleading. Instead, we like to think about starting with the first term in the first set

Notes:

of parentheses and multiplying it by the first term of the second set, then the second term of the second set, then the third term of the second set, etc. Then, we move to the second term in the first set, and do the same thing. Let's see this in action:

Example 22   **Expanding the Product of Quadratic Functions**
Expand $g(t)h(t)$ for $g(t) = 2t^2 + 3t + 4$ and $h(t) = t^2 - t - 3$.

Solution   Like before, we need to make sure to put parentheses around each of the functions before we multiply; this gives us:

$$\begin{aligned}
g(t)h(t) &= (2t^2 + 3t + 4)(t^2 - t - 3) \\
&= 2t^2(t^2 - t - 3) + 3t(t^2 - t - 3) + 4(t^2 - t - 3) \\
&= 2t^4 - 2t^3 - 6t^2 + 3t^3 - 3t^2 - 9t + 4t^2 - 4t - 12 \\
&= 2t^4 + t^3 - 5t^2 - 13t - 12
\end{aligned}$$

(1.29)

Here we had a fair bit of combining of like-terms to take care of after we finished multiplying; there was one t^4 term, two t^3 terms, three t^2 terms, two t terms, and one constant term. After combining the like-terms, we get

$$g(t)h(t) = 2t^4 + t^3 - 5t^2 - 13t - 12$$

In each of our examples so far, we've only worked with two sets of parentheses. We can expand on this process to work in situations where we have three or more sets of parentheses. Personally, we like working from left to right, so we start by expanding the first two sets of parentheses. Then, we take that result and expand it with the next set. We continue until everything has been expanded. We make sure to combine like terms as part of each expansion because otherwise the numbers of terms gets really big, really fast. As we saw in our expansion of quadratics, we had nine terms before we combined like-terms; after combining, we only had five.

Example 23   **Expanding with Three Set of Parentheses**
Expand $f(x)g(x)h(x)$ where $f(x) = x - 4$, $g(x) = -x + 3$, and $h(x) = 2x + 1$,

Solution   We'll work left right; we'll expand the first two sets of paren-

Notes:

theses and then that result with the third set.

$$f(x)g(x)h(x) = (x-4)(-x+3)(2x+1)$$
$$= (-x^2 + 3x + 4x - 12)(2x+1)$$
$$= (-x^2 + 7x - 12)(2x+1) \quad\quad (1.30)$$
$$= -2x^3 - x^2 + 14x^2 + 7x - 24x - 12$$
$$= -2x^3 + 13x^2 - 17x - 12$$

Notice that we did not show every single step of the process here. Realisti-cally, this is the level of detail you would typically see on this type of problem. Until you are fully confident with the process we do recommend showing every step, but once you are comfortable with the ideas, you can show work like we did in this problem. Notice that we did combine any like terms after the first distribution step, and then again at the very end, giving us a final answer of

$$\boxed{f(x)g(x)h(x) = -2x^3 + 13x^2 - 17x - 12}$$

These methods will work, no matter how many sets of parentheses you are working with and no matter how many terms are in each set. There are a few other situations where we will need to use these techniques that may not be obvious. For example, if we have $f(x) = x + 3$ and $g(x) = x^2$, we know that we could combine these two functions in many ways. If we do the composition of g with f, we would have $g(f(x)) = (x+3)^2$. We could leave the function in this form, but there may be situations where we want to expand it. Here we would need to remember that $(x+3)^2$ is the same as $(x+3)(x+3)$, since squaring means we should multiply the term by itself. Similarly, if we have $h(x) = x^3$ and want to find the composition of h with f, we would have $h(f(x)) = (x+3)^3$, or $h(f(x)) = (x+3)(x+3)(x+3)$. Be careful in these situations to work one step at a time; many students are tempted to write things like $(x+3)^2 = x^2 + 3^2$ or $(x+3)^3 = x^3 + 3^3$, but through the process of expanding, you can see that this shortcut is no good because it gives us a false statement.

Common Expansion Patterns

There are some expansions that show up very frequently in mathematics. You may find it useful to memorize some of these patterns, however be sure to expand each by hand at least once so that you can see and understand why these patterns are correct. Here are three expansions that you may see frequently:

- $(a+b)^2 = a^2 + 2ab + b^2$

Notes:

- $(a+b)(a-b) = a^2 - b^2$
- $(a+b)^3 = a^3 + 3a^2b + 3ab^2 + b^3$

In each of these patterns, a and b can be anything. Let's take a look at working with one of these patterns when a and b are a bit complicated.

Example 24   Expanding Using Expansion Rules
Using the expansion rules given above, expand $(2xy - 3xyz)^2$

Solution   Since here we are squaring, the closest form is the the first rule, $(a+b)^2 = a^2 + 2ab + b^2$. However, in the first rule, the terms are added, and in our problem the second term is subtracted. However, we can fix this by writing $(2xy + (-3xyz))^2$ instead of $(2xy - 3xyz)^2$, since a and b can be anything, positive or negative. Next, we need to identify what we should use for a and for b. In the rule, the first term is a, and our first term is $2xy$, so we should use $a = 2xy$. The second term in the pattern is b, and our second term is $-3xyz$, so we should use $b = -3xyz$. Notice that our b includes the negative. Now that we've identify the rule we need and every part of the rule, we can complete our expansion. As we do this, we will put parentheses around each a and each b to make sure everything gets used correctly.

$$
\begin{aligned}
(2xy - 3xyz)^2 &= (2xy + (-3xyz))^2 \\
&= (2xy)^2 + 2(2xy)(-3xyz) + (-3xyz)^2 \\
&= (2xy)(2xy) + 2(2xy)(-3xyz) + (-3xyz)(-3xyz) \\
&= 4x^2y^2 - 12x^2y^2z + 9x^2y^2z^2
\end{aligned}
\tag{1.31}
$$

Notice that we were very careful in the places where we were working with negative signs. A common mistake is to leave off a negative sign, but this can drastically change your final answer. Here, we can't combine any terms because the exponents on z are different for each of the terms, so our final answer is

$$
\boxed{(2xy - 3xyz)^2 = 4x^2y^2 - 12x^2y^2z + 9x^2y^2z^2}
$$

Factoring

As mentioned at the start of this section, expanding and factoring are inverse actions; expanding moves us from the product of polynomials to a single, expanded, polynomial, and factoring moves us from that single expanded polynomial back to the product of polynomials. We move back and forth between

Notes:

the two forms because sometimes one form is much more useful than another. Expanding often comes in handy in calculus when you are taking derivatives or evaluating an integral, and factoring can be used to simplify rational functions (functions that are the quotient of polynomials) and to determine where a function is equal to zero (also know as finding its roots). Mostly we will work on factoring quadratic and cubic functions; higher degree functions can be very difficult to factor and are only rarely need to be factored in calculus. Also, we will only look at examples where there is no obvious factor that is shared by all terms; for example, $h(t) = 2t^3 + 14t^2 + 20t$ has $2t$ as a factor for each term, so the first step would be to factor out the $2t$. This would give $h(t) = (2t)(t^2 + 14t + 20)$. Your first step in factoring should always be to look for common factors and deal with those first. In this section, we will discuss how to find the less-obvious factors.

In order to factor, it is important to be comfortable with expanding since they are inverse actions. We'll start by looking at how to factor a quadratic function where the leading term, x^2 has a coefficient of one. Quadratics can't always be factored (we'll get back to this later), but quadratics of this form are the easiest to work with. When we factor a quadratic, we will end up with the product of two linear functions, called *factors*, if it is possible to factor the quadratic. For higher degree polynomials, our factors may be linear or quadratic. A polynomial can only have as many linear factors as its degree, so a cubic can have at most three linear factors, and a fourth degree polynomial can have a most four linear factors, Let's take a quick look at what the product of two linear function looks like:

$$\begin{aligned} (x + a)(x + b) &= x^2 + bx + ax + ab \\ &= x^2 + (a + b)x + ab \end{aligned} \tag{1.32}$$

Here, we are only looking at situations where a and b are both integers. They can be positive, negative, or zero. Let's identify some key characteristics of equation 1.32. We see that in each linear term, x has a coefficient of one. In the expanded form, the x^2 comes from multiplying the two x terms together, so it also has a coefficient of one. In the expanded form, the constant term is a product of a and b, and the x term's coefficient is $a + b$. The combination of these facts will help us factor quadratics. We know that if we look at the constant term in the expanded version, it will be the product of the constants from the linear terms. This will give us a good starting point to look for factors. We can then limit the possibilities some by looking at the x term in the expanded form. Its coefficient is the sum of these two constants. For example, if we have $x^2 + 5x + 4$, we have several pairs of integers that could be multiplied together to give us 4. Let's look at how we can eliminate some of these pairs:

Notes:

Example 25   Factoring a Quadratic
Factor the quadratic function $f(x) = x^2 + 5x + 4$.

Solution   As we noted above, the best starting point is to look for pairs of integers that we can multiply to get the constant term. We know that to get 4, we could multiply any of the following pairs to get 4:

(A) 4 and 1

(B) 2 and 2

(C) -4 and -1

(D) -2 and -2

Now, we'll look at the x term in the quadratic. It has a coefficient of 5, so we need to figure out which pair of numbers will add up to 5. We can pretty quickly see that the only pair that can add up to 5 is 4 and 1. That tells us that the factors of x^2+5x+4 are $x+4$ and $x+1$. We often like to verify that we factored correctly by multiplying and expanding the factors. Let's check:

$$(x + 4)(x + 1) = x^2 + x + 4x + 4$$
$$= x^2 + 5x + 4$$
$$= f(x)$$

This verifies that our factors are correct, and verifies that our final answer is

$$\boxed{x^2 + 5x + 4 = (x + 4)(x + 1)}$$

Let's look at a few more examples so that we can compare them and look for some patterns that might help us factor more quickly.

Example 26   Factoring a Quadratic
Factor the quadratic function $g(x) = x^2 - 5x + 6$.

Solution   We'll start like we did in our last example by looking for pairs of integers that multiply to give us 6:

(A) 6 and 1

(B) 3 and 2

(C) -6 and -1

(D) -3 and -2

Out of these pairs, only -3 and -2 add to give us -5, the x coefficient in the quadratic. This tells us that the factors are $x - 3$ and $x - 2$. So, we have

Notes:

$$x^2 - 5x + 6 = (x - 3)(x - 2)$$

Example 27   **Factoring a Quadratic**

Factor the quadratic function $h(x) = x^2 - 7x - 18$.

   Solution   We'll start like we did in our last example by looking for pairs of integers that multiply to give us -18:

(A) -18 and 1

(B) -9 and 2

(C) -6 and 3

(D) -3 and 6

(E) -2 and 9

(F) -1 and 18

Out of these pairs, only -9 and 2 add to give us -7, the x coefficient in the quadratic. This tells us that the factors are $x - 9$ and $x + 2$. So, we have

$$x^2 - 7x - 18 = (x - 9)(x + 2)$$

Example 28   **Factoring a Quadratic**

Factor the quadratic function $m(x) = x^2 + 3x - 18$.

   Solution   We'll start like we did in our last example by looking for pairs of integers that multiply to give us -18:

(A) -18 and 1

(B) -9 and 2

(C) -6 and 3

(D) -3 and 6

(E) -2 and 9

(F) -1 and 18

Out of these pairs, only -3 and 6 add to give us 3, the x coefficient in the quadratic. This tells us that the factors are $x - 3$ and $x + 6$. So, we have

$$x^2 + 3x - 18 = (x - 3)(x + 6)$$

Notes:

Notice that in examples 25 and 26, the constant term in the quadratic is positive. This tells us that our integers in our pairs both need to have the same sign. What about in examples 27 and 28? What relationship does the constant in the quadratic have with the integers in our pairs? In examples 27 and 28, is there anything about the coefficient on x in the quadratic that relates to the signs of the integers in the pairs? You may find identifying some of these patterns useful, and it will help you understand the ideas in factoring more deeply. Don't feel like you have to memorize these patterns; it's much better to be comfortable with the process of factoring than to remember rules that you don't understand the application of.

Factors and Roots

When we find the factors of a polynomial, we are only a couple of steps away from finding the roots of the function. The *roots* are the inputs of the function that have zero for their output. For example, $f(x) = x^2 + 5x + 4$ from example 25 has $x = -1$ and $x = -4$ as roots because $f(-1) = 0$ and $f(-4) = 0$. We saw in example 25 that the factors of $f(x)$ are $x + 1$ and $x + 4$. We can use these factors to find the roots and vice versa. If we set each factor equal to 0 and solve for the input variable, x, we will get the roots of the function:

$$x + 1 = 0 \qquad\qquad x + 4 = 0$$
$$x = -1 \qquad\qquad x = -4$$

Similarly, we can go backwards and find the factors from the roots:

$$x = -1 \qquad\qquad x = -4$$
$$x + 1 = 0 \qquad\qquad x + 4 = 0$$

This relationship between factors and roots is quite handy because there is a nice formula that will help us determine the roots of a quadratic: the quadratic formula. The quadratic formula tells us that the roots of the function $f(x) = ax^2 + bx + c$ are:

$$x = \frac{-b \pm \sqrt{b^2 - 4ac}}{2a} \tag{1.33}$$

The $\pm$ sign tells us that we will have two roots; one root we find by using $+$ and the second root we find by using $-$. This is a more compact way of expressing the formula for the roots, rather than writing it as two separate formulas. The quadratic function is quite useful when we can't easily find the factors like in our earlier examples. Problems will rarely tell you that you need to use the quadratic formula; it is up to you to make that decision. In fact, some people prefer using

Notes:

the quadratic formula all the time rather than factoring like we did earlier. Let's take a look at its use.

Example 29   Factoring a Quadratic

Factor the quadratic function $h(t) = 6t^2 - 7t + 2$.

   Solution   Here, our function has t instead of x, but it really is in the form we need to use the quadratic formula; we'll just make sure to give the answer with t instead of x. We'll find our roots, and then use those to help us find our factors. We'll start by identifying the values for a, b, and c, and then plugging them into the quadratic formula. The coefficient on t^2 is 6, so that tells us $a = 6$. The coefficient on t is -7, so that tells us $b = -7$. Lastly, the constant is 2, so $c = 2$. We'll take these values and plug them into our formula:

$$t = \frac{-(-7) \pm \sqrt{(-7)^2 - 4(6)(2)}}{2(6)}$$

$$= \frac{-(-7) \pm \sqrt{49 - 48}}{2(6)}$$

$$= \frac{-(-7) \pm \sqrt{1}}{12}$$

$$= \frac{7 \pm 1}{12}$$

From here, we'll split into two formulas so we get both roots:

$$t = \frac{7+1}{12} = \frac{8}{12} = \frac{2}{3}$$

$$t = \frac{7-1}{12} = \frac{6}{12} = \frac{1}{2}$$

This tells us our two roots: $\frac{1}{2}$ and $\frac{2}{3}$. If we rewrite to find our factors, we get $t - \frac{1}{2}$ and $t - \frac{2}{3}$ as factors. However, These are not quite enough. If we multiply them out, t^2 only has a coefficient of 1, not 6, like in $h(t)$. This tells us that we also have 6 as a factor, so our final answer is

$$\boxed{h(t) = 6\left(t - \frac{1}{2}\right)\left(t - \frac{2}{3}\right)}$$

   Many people might not like this final answer and may factor slightly differently. We could rewrite this answer a bit:

Notes:

$$h(t) = 6\left(t - \frac{1}{2}\right)\left(t - \frac{2}{3}\right)$$

$$= 2 \times 3 \times \left(t - \frac{1}{2}\right)\left(t - \frac{2}{3}\right)$$

$$= \left[2\left(t - \frac{1}{2}\right)\right]\left[3\left(t - \frac{2}{3}\right)\right]$$

$$= (2t - 1)(3t - 2)$$

Both of these answers are equally valid; some prefer the second form because there are no fractions. Some prefer the first form because it's easier to identify the roots, $t = \frac{1}{2}$ and $t = \frac{2}{3}$. Either way, the function $h(t)$ is considered factored.

Irreducible Quadratics

As we mentioned earlier, not all quadratics can be factored. If a quadratic cannot be factored, we say it is *irreducible*, meaning it can't be "reduced" into the product of linear functions. These quadratics can be identified through use of the quadratic formula. If a quadratic is irreducible, we'll run into a problem using the quadratic formula. The *discriminant*, the part under the square root, will be negative. This will tell us that the function has no real number roots, only a pair of imaginary roots. If you are factoring a polynomial and run into an irreducible quadratic, just leave it alone. The irreducible quadratic would be considered one of the factors of the polynomial.

Factoring Cubic Functions

Factoring cubic functions can be a bit tricky. There is a special formula for finding the roots of a cubic function, but it is very long and complicated. In fact, it very rarely gets used. Instead, mathematicians build off of the ideas we've already learned this section. Typically, the first place to start with a cubic function is by finding one of the roots. To do this, we start by listing all integer factors of the constant term. Then, we plug each of these factors into the function to see if any of them are roots. We start by trying the numbers that are easiest to work with like 1, -1, and other small integers. Once we find one root, we'll stop plugging in factors of the constant term because we know we've found one factor of the polynomial. Before we worry about the next step of the process, let's see this first step.

Example 30   **Finding a Factor of a Cubic Function**
Find a factor of the function $f(x) = x^3 + 8x^2 + 21x + 18$.

Notes:

Solution   Here, the constant term of the cubic is 18, so we'll start by listing all of its factors, positive and negative. The factors are: 18, 9, 6, 3, 2, 1, -1, -2, -3, -6, -9, and -18. There are a bunch, so as mentioned above, we'll start by checking the "easy" numbers to see if any of them are roots.

- $f(1) = (1)^3 + 8(1)^2 + 21(1) + 18 = 1 + 8 + 21 + 18 \neq 0$
- $f(-1) = (-1)^3 + 8(-1)^2 + 21(-1) + 18 = -1 + 8 - 21 + 18 \neq 0$
- $f(-2) = (-2)^3 + 8(-2)^2 + 21(-2) + 18 = -8 + 32 - 42 + 18 = 0$

Since $f(-2) = 0$, we know that $x = -2$ is a root of $f(x)$, telling us that

$$\boxed{x + 2 \text{ is a factor of } f(x)}$$

Notice that in the previous example, we started with the easy numbers first. Also, we can eliminate half of these factors pretty quickly. In $f(x)$, every term has a positive coefficient. We know that if x is positive, x^3 and x^2 are also positive, and we can't add up a bunch of positive numbers and get 0, so we don't need to check any of the positive factors, only the negative factors.

This gives us one factor, but it doesn't help us fully factor this polynomial. We could try looking for other roots, but we already know that it's possible to have an irreducible quadratic as a factor, or even just a quadratic that doesn't have integer roots. The most reliable method of finding other the other factors of a cubic is with *polynomial long division*. Polynomial long division works similarly to regular long division with numbers. We'll finish factoring $f(x) = x^3 + 8x^2 + 21x + 18$ as an example, and we'll describe each step in detail. First, we want to start with the same kind of set up we use for long division, but this time we will be dividing $x^3 + 8x^2 + 21x + 18$ by $x + 2$, the factor we already found.

Example 31   **Factoring a Cubic Function**
Completely factor the function $f(x) = x^3 + 8x^2 + 21x + 18$.

Solution   In example 30, we found that $x = -2$ is a root of $f(x)$, telling us that $x + 2$ is a factor of $f(x)$. Now that we have one factor, we can use polynomial long division to help find the remaining factors. We'll start by setting up the polynomial long division. The initial setup is just like long division with numbers:

$$x + 2 \overline{\smash{)}x^3 + 8x^2 + 21x + 18}$$

With polynomial long division, we will focus on the highest power of x at each step. Initially, the highest power term of our dividend, $x^3 + 8x^2 + 21x + 18$

Notes:

is x^3, and the highest power term of the divisor, $x + 2$, is x. If we divide x^3 by x, we get x^2. This will go above the line, and we will subtract $x^2(x + 2) = x^3 + 2x^2$ from the dividend:

$$
\begin{array}{r}
x^2 \phantom{{}+ 21x + 18} \\
x + 2\overline{\smash{\big)}\ x^3 + 8x^2 + 21x + 18} \\
\underline{-(x^3 + 2x^2)} \\
6x^2 + 21x
\end{array}
$$

When we do the subtraction, we are left with $6x^2$, and we bring down the next highest power term from the dividend, $21x$. Again, we will only look at the highest power terms, $6x^2$, and x. If we divided $6x^2$ by x, we get $6x$. This goes above the line, and we will subtract $6x(x + 2) = 6x^2 + 12x$ from what's left of the dividend:

$$
\begin{array}{r}
x^2 + 6x \phantom{{}+ 18} \\
x + 2\overline{\smash{\big)}\ x^3 + 8x^2 + 21x + 18} \\
\underline{-(x^3 + 2x^2)} \\
6x^2 + 21x \\
\underline{-(6x^2 + 12x)} \\
9x + 18
\end{array}
$$

This subtraction leaves us with $9x$, and we bring down the last term from our dividend, 18. Looking at the highest power terms, we have $9x$ and x. If we divided $9x$ by x, we get 9. This goes above the line, and we will subtract $9(x+2) = 9x+18$ from what's left of the dividend:

$$
\begin{array}{r}
x^2 + 6x + 9 \\
x + 2\overline{\smash{\big)}\ x^3 + 8x^2 + 21x + 18} \\
\underline{-(x^3 + 2x^2)} \\
6x^2 + 21x \\
\underline{-(6x^2 + 12x)} \\
9x + 18 \\
\underline{-(9x + 18)} \\
0
\end{array}
$$

After we complete the subtraction, we get 0 and we have no other terms left from our dividend. This means we are done with the polynomial long division and have no remainder. The lack of remainder verifies that $x + 2$ is a factor of $f(x)$; if there were a remainder, it would not be a factor. So far, we have that $f(x) = (x + 2)(x^2 + 6x + 9)$.

We're not quite done yet, because we have a quadratic term, and we haven't checked to see if we can factor it or if it's irreducible. We'll try to factor it first. We see that the constant term is 9; our factor pairs of 9 are: 9 and 1, 3 and 3, -3 and -3, and -9 and -1. The pair 3 and 3 adds to 6, so we see that $x^2 + 6x + 9 =$

Notes:

$(x + 3)(x + 3)$. We can condense this a little bit by writing $(x + 3)^2$ instead of $(x + 3)(x + 3)$.

Altogether, we have that

$$f(x) = (x + 2)(x + 3)^2$$

We won't show how to factor polynomial with a degree higher than 3, but the process is very similar. You would start by trying to find a root; once you find a root you can rewrite to get a factor and you can do polynomial long division. The polynomial long division will tell you a second factor. Keep repeating those steps until you only have quadratic and linear factors.

Factoring by Grouping

In some special circumstances, we can use a different method for factoring cubics, called factoring by grouping. With this method, we will "group" the x^3 and x^2 terms together and factor out any common terms, and we will "group" the x and the constant terms together and factor any common terms from those. This method is fast and efficient when it works, but it does not always work. We'll look at an example where it does work and one where it doesn't.

Example 32   Factoring a Cubic Function
Completely factor $f(t) = t^3 + t^2 - 4t - 4$.

Solution   We will start by grouping and then factoring each group:

$$f(t) = t^3 + t^2 - 4t - 4 = (t^3 + t^2) + (-4t - 4)$$
$$= t^2(t + 1) + (-4)(t + 1)$$
$$= t^2(t + 1) - 4(t + 1)$$
$$= (t^2 - 4)(t + 1)$$

Notice that after factoring each group, we were left with $t + 1$ for each. This means that $t + 1$ is a factor of $f(t)$. What we factored out, t^2 and -4, combine to give us another factor, $t^2 - 4$. This is a quadratic, so we need to see if it can be factored. The factor pairs of the constant, -4, are -4 and 1, -2 and 2, and -1 and 4. We see that -2 and 2 add to zero, telling us that the factors of $t^2 - 4$ are $t - 2$ and $t + 2$. In total, we get

$$f(t) = (t + 1)(t - 2)(t + 2)$$

Notes:

To see an example where factoring by grouping doesn't work, let's look back at the function from Example 31, $f(x) = x^3 + 8x^2 + 21x + 18$. We can start by grouping and factoring each group:

$$f(x) = x^3 + 8x^2 + 21x + 18 = (x^3 + 8x^2) + (21x + 18)$$
$$= x^2(x + 8) + 3(7x + 6)$$

Notice that what's left after pulling out the common factors are $x + 8$ and $7x + 6$. These are two very different terms, and neither looks like any of the factors we found earlier: $x + 2$ and $x + 3$. Since these are different, we are not able to find any factors of $f(x)$ this way. It can be a good idea to try out factoring by grouping before diving into the polynomial long division method, but factoring by grouping is not guaranteed to help you find the factors of your function. As we saw in Example 32, when factoring by grouping works, it works well and is very quick, but it's downfall is that it is not guaranteed to find a factor.

Common Patterns

In our section on expanding, we saw some common patterns that can be used as shortcuts when expanding certain forms. These patterns work equally well in the opposite direction; if we see something that fits one of the expanded forms, we'll know from the pattern what the factors are. Here are those patterns, and a few others, written with the expanded form first.

- $a^2 + 2ab + b^2 = (a + b)^2$

- $a^2 - b^2 = (a + b)(a - b)$

- $a^3 + 3a^2b + 3ab^2 + b^3 = (a + b)^3$

- $a^3 + b^3 = (a + b)(a^2 - ab + b^2)$

- $a^3 - b^3 = (a - b)(a^2 + ab + b^2)$

It's good to note here that a and b can be anything, integers or decimals, positive or negative, and they can include variables. For example, after completing the polynomial division, we ended up with $f(x) = (x + 2)(x^2 + 6x + 9)$. The quadratic fits one of our patterns: if we let $a = x$ and $b = 3$, it fits the first pattern. This pattern then tells us that $x^2 + 6x + 9 = (x + 3)^3$, which we were able to find with our earlier methods. Memorizing these patterns can be useful and save some time, but it's much more important to be comfortable with the

Notes:

other methods we discussed. Let's take a look at an example of using a pattern where a and b are a bit more complicated.

Example 33 **Factoring Using Patterns**

Factor $g(t) = 8t^3 - \frac{1}{27}$ completely.

   Solution   Here we see that $g(t)$ has only two terms, and that one of them has t^3. This points me towards the last two patterns: they both have parts raised to the third power and only have two terms each. With $g(t)$, we have subtraction, not addition, so this points us to the last rule. We need to figure out what a and b could be. The pattern starts with a^3 and $g(t)$ starts with $8t^3$, so it looks like we have $a^3 = 8t^3$. This works out if $a = 2t$ since $(2t)^3 = 8t^3$.

Next, we need to figure out b. The second term in the pattern is b^3 and the second term in $g(t)$ is $\frac{1}{27}$. This tells us that $b^3 = \frac{1}{27}$, or that $b = \frac{1}{3}$.

Now, we can use the pattern:

$$
\begin{aligned}
g(t) &= 8t^3 - \frac{1}{27} \\
&= (2t)^3 - \left(\frac{1}{3}\right)^3 \\
&= \left[2t - \frac{1}{3}\right]\left[(2t)^2 + (2t)\left(\frac{1}{3}\right) + \left(\frac{1}{3}\right)^2\right] \\
&= \left(2t - \frac{1}{3}\right)\left(4t^2 + \frac{2t}{3} + \frac{1}{9}\right)
\end{aligned}
$$

We don't need to go any further. Normally, we would check to see if the quadratic can be factored or if it is irreducible, but the patterns have all been fully reduced, meaning that we will never be able to factor the quadratic in this pattern.

Notes:

Exercises 1.3

Terms and Concepts

1. In your own words, explain the relationship between factors and roots.

2. If $x = 2, x = 5$, and $x = -1$ are the only roots of the function $f(x)$, what are the factors of $f(x)$?

3. What does it mean for a quadratic to be irreducible?

4. What is the maximum number of linear factors that $g(t) = t^6 + t^4 - 2t^2 + 1$ could have?

5. What is the maximum number of roots that $g(t) = t^6 + t^4 - 2t^2 + 1$ could have?

Problems

In exercises 6 – 12, expand and simplify the given expressions.

6. $3a(2b + 5)(a - 2b)$

7. $(2t + 7)^2$

8. $2(x^2 + 3x + 4)(2x + 3)$

9. $(t^2 - 3t + 1)(2t) - (t^2 + 2)(2t - 3)$

10. $(x^3 + x - 2)(2) - (2x + 1)(3x^2 + 1)$

11. $(x^2 + 3x - 1)(4x) + (2x^2 - 5)(2x + 3)$

12. $3(\theta^2 + 4)^2(2\theta)$

In exercises 13 – 20, factor each function completely.

13. $g(x) = 4x^2 + 4x + 1$

14. $y(z) = z^2 - 7z + 10$

15. $f(k) = k^4 - 27k$

16. $\theta(\gamma) = 6\gamma^2 - \gamma - 2$

17. $x(z) = 3z^3 + 6z^2 - 24z$

18. $y(x) = x^3 + 8$

19. $f(x) = 2x^3 - x^2 - 5x - 2$

20. $f(y) = y^3 - 5y^2 - 2y + 24$

In exercises 21 – 24, determine the difference quotient of the given function.

21. $g(t) = t^3 + 1$

22. $y(x) = 2x^2 - 5$

23. $f(x) = x^3 + x^2 - x$

24. $g(x) = 4x^2 + 2x$

In exercises 25 – 27, find all real roots of the given function.

25. $g(x) = 4x^2 + 2x$

26. $f(x) = x^3 + x^2 - x$

27. $y(x) = 4x^2 - 5$

In exercises 28 – 30, factor the given function, and relate the factors with the roots found in exercises 25 -27.

28. $g(x) = 4x^2 + 2x$

29. $f(x) = x^3 + x^2 - x$

30. $y(x) = 4x^2 - 5$

1.4   Radicals and Exponents

In this section, we will look at properties of exponents. Here, these rules apply to any type of function that involves exponents, namely power functions and exponential functions. However, this section will mostly focus on power functions, functions where the base is the variable and the exponent is a constant. We'll discuss several exponent rules, show you how to use them, and explain the reasoning behind these rules.

> In this section, we are assuming that all variables are strictly positive, meaning that they cannot be negative nor can they be zero.

This is to ensure that we won't run into any issues with dividing by zero or trying to take a square root of a negative number. Later, when we discuss functions domains, we will revisit these problems and explain how to deal with general variables and not just variables that are strictly positive. Before we dive into the different rules, we need to have a solid understanding of what exponents are and what they mean.

When we first start learning math, we often start with addition. We then quickly see that repeated addition can be useful. We run into problems like: "You have 4 dogs and want to give each dog 3 treats. How many treats do you need?" We solve these with repeated addition: $3 + 3 + 3 + 3 = 12$, or three treats for each of the four dogs. We then learn that repeated addition happens often, so we develop a new notation, multiplication. For our example problem, we would do 3×4 to say we need to add three, four times. Exponents take this one step further. When we need to do repeated multiplication, like if we need to find the volume of a cube, we can shorten the notation by using exponents. To find our volume, we would multiply the side length by itself three times to get length times width times height, but with a cube these lengths are all the same. We can write $V(x) = (x)(x)(x) = x^3$. Here, the exponent tells us how many times to do the multiplication.

The first exponent rule we will examine is

$$x^a x^b = x^{a+b} \tag{1.34}$$

Here, the first term, x^a tells us to multiply x by itself a times and the second term tells us to multiply it b times. Together, that says we need to multiply x a total of $a + b$ times, giving us x^{a+b}. As an example, $x^2 x^3 = x^{2+3} = x^5$.

Next, let's look at

$$x^{-a} = \frac{1}{x^a} \tag{1.35}$$

Notes:

This rule build off of our last rule. If we have $x^5 x^{-2}$, rule 1.34 tells us we really have $x^{5+(-2)} = x^{5-2} = x^3$. We went from having x multiplied 5 times to having x multiplied only 3 terms, meaning we have removed two of the multiplications. We removed a multiplication through a division:

$$\frac{x^5}{x^2} = \frac{(x)(x)(x)(x)(x)}{(x)(x)} = (x)(x)(x) = x^3$$

This shows us that a negative exponent tells us we have division rather than multiplication. We can also combine this rule with some of our rules from fractions. If we have $\frac{1}{x^{-a}}$, we can start by replacing x^{-a} with $\frac{1}{x^a}$. This gives us

$$\frac{1}{x^{-a}} = \frac{1}{\frac{1}{x^a}}$$

From our fractions rules, we know that dividing by a fraction is the same as multiplying by its reciprocal, so we have

$$\frac{1}{x^{-a}} = \frac{1}{\frac{1}{x^a}}$$
$$= 1 \times \frac{x^a}{1}$$
$$= x^a$$

The third rule we will discuss is

$$(x^a)^b = x^{ab} \tag{1.36}$$

This rule builds directly off of our first rule as well. $(x^a)^b$ tells us we need to multiply x^a by itself b times. Since x^a multiplies x by itself a times, $(x^a)^b$ tells us to multiply x by itself a total of ab times. For example, $(x^2)^3 = (x^2)(x^2)(x^2) = x^{2+2+2} = x^{(2)(3)} = x^6$. We can also use this rule when there is a product or quotient inside the parentheses, but not if there is an addition or subtraction. For example, we can say that $(x^2 y^3)^2 = (x^2)^2(y^3)^2 = x^4 y^6$, and that $\left(\frac{x^2}{y^3}\right)^2 = \frac{(x^2)^2}{(y^3)^2} = \frac{x^4}{y^6}$, but we cannot apply this rule to $(x^2 + y^3)^2$. Here, we would need to rewrite as $(x^2 + y^3)(x^2 + y^3)$ and distribute as we saw in our previous section on expanding.

Our last rule focuses on the inverse function, or how to "undo" an exponent. We've seen these functions before. These are our root functions. A square root "undoes" squaring and a cube root "undoes" cubing. In general, we have

Notes:

$$(x^a)^{1/a} = x \qquad (1.37)$$

and

$$(x^{1/a})^a = x \qquad (1.38)$$

Both of these come from rule 1.36. Additionally, you might see $x^{1/a}$ written as $\sqrt[a]{x}$. Mathematicians call $\sqrt[a]{x}$ the *radical* form and $x^{1/a}$ the *exponential* form. Both of these have the same meaning, they just look a bit different. Anytime you see $\sqrt[a]{x}$, you can replace it with $x^{1/a}$ and vice versa.

These rules will all be quite handy in calculus. In both integral and differential calculus, we will have rules that work well when we have a power function, but won't work for other forms of functions. By being able to rewrite functions like $f(x) = \frac{1}{x^2}$, as power functions ($f(x) = x^{-2}$ here), other calculations will be simplified. Our rules are summarized below.

Exponent Rules

- $x^a x^b = x^{a+b}$

- $\dfrac{1}{x^{-a}} = x^a$

- $(x^a)^b = x^{ab}$

- $x^{-a} = \dfrac{1}{x^a}$

- $x^{1/a} = \sqrt[a]{x}$

Let's look at a few examples of working with exponent rules.

Example 34   Simplifying Exponents

Simplify $\left(\dfrac{x^2 y^4}{x\sqrt{y}} \right)^2$

Solution   Anytime we simplify, we need to remember our order of operations. The order of operations tells us to start with terms that are inside of parentheses, so we will work on simplifying the fraction before we worry about the exponent on the outside. First, we will write everything using exponents rather than radicals so we can use our exponent rules more easily in the rest of the problem.

$$\left(\frac{x^2 y^4}{x\sqrt{y}} \right)^2 = \left(\frac{x^2 y^4}{x y^{1/2}} \right)^2$$

Next, we will eliminate the fraction by using negative exponents on the terms that are in the denominator. After rewriting, we will combine any like terms.

Notes:

$$\left(\frac{x^2y^4}{xy^{1/2}}\right)^2 = \left(x^2y^4x^{-1}y^{-1/2}\right)^2$$

$$= \left(x^2x^{-1}y^4y^{-1/2}\right)^2$$

$$= \left(x^{2-1}y^{4-1/2}\right)^2$$

$$= \left(x^1y^{8/2-1/2}\right)^2$$

$$= \left(xy^{7/2}\right)^2$$

Now that everything inside the parentheses is simplified as much as possible, we will use our third exponent rule to finish simplifying. Our third rule says that $(x^a)^b = x^{ab}$. We need to make sure we distribute the exponent that is outside of the parentheses to each term inside of the parentheses. This give us

$$\left(xy^{7/2}\right)^2 = (x)^2(y^{7/2})^2$$

$$= x^2y^{14/2}$$

$$= x^2y^7$$

So, in the end, we get that

$$\boxed{\left(\frac{x^2y^4}{x\sqrt{y}}\right)^2 = x^2y^7}$$

Example 35   **Simplifying Exponents**

Simplify $\left(\sqrt{y} + \sqrt{x}\right)^2$.

Solution   We'll start again by focusing on the terms inside the parentheses and rewriting all radicals as exponents. This gives us

$$\left(\sqrt{y} + \sqrt{x}\right)^2 = \left(y^{1/2} + x^{1/2}\right)^2$$

There is nothing that we can simplify inside the parentheses, so we now need to apply the exponent on the outside of the parentheses. Inside the parentheses

Notes:

we have two terms that are added, so we can't apply an exponent rule here. We will need to rewrite and then expand.

$$\left(y^{1/2} + x^{1/2}\right)^2 = \left(y^{1/2} + x^{1/2}\right)\left(y^{1/2} + x^{1/2}\right)$$
$$= (y^{1/2})^2 + 2y^{1/2}x^{1/2} + (x^{1/2})^2$$
$$= y + 2y^{1/2}x^{1/2} + x$$

We don't have any like terms, so we can't simplify any further. We could rewrite slightly, but this is a matter of personal preference. We have three other ways we could write this final answer. We could use exponent rules to rewrite the middle term since $y^{1/2}x^{1/2} = (yx)^{1/2}$, giving us $y + 2(yx)^{1/2} + x$. We could also use radicals and write either $y + 2\sqrt{y}\sqrt{x} + x$ or $y + 2\sqrt{yx} + x$. All of these four answers are fully simplified, and are equally valid. Probably the most common form is

$$\boxed{\left(\sqrt{y} + \sqrt{x}\right)^2 = y + 2\sqrt{yx} + x}$$

Many people struggle with evaluating radicals or fractional exponents by hand. Let's take a look at how we can evaluate these types of terms.

Example 36　　**Evaluating Radicals**

Evaluate $8^{2/3}$.

Solution　　As first glance, this looks like we won't be able to do much with it. However, we can use our exponent rules to help us evaluate it. We can rewrite this as $(8^2)^{1/3}$ or as $(8^{1/3})^2$. We prefer the second version. With the first version we would have $(8^2)^{1/3} = (64)^{1/3}$, but this is tricky to deal with by hand because not many people have perfect cubes memorized, so we would need to factor 64.

If we use the second version, $(8^{1/3})^2$, we would start by finding the cube root of 8. When we factor, we get $8 = 2 \times 2 \times 2$, which show us that $2 = 8^{1/3}$. This gives us $(8^{1/3})^2 = (2)^2 = 4$, so our final answer is

$$\boxed{8^{2/3} = 4}$$

Notes:

Exercises 1.4

Terms and Concepts

1. Do exponent rules apply to root functions? Explain.

2. Explain why a negative exponents moves the term to the denominator and gives it a positive exponent.

3. Is $3x(2x + 3)^{-5/3}$ in radical or exponential form?

4. Is $\dfrac{3x}{\sqrt[3]{(2x + 3)^5}}$ in radical or exponential form?

Problems

In exercises 5 – 7, write the given term without using exponents.

5. $(8x_1 - 5x_2 + 11)^{-1/3}$

6. $(-2x + y)^{-1/5}$

7. $(5x - 2)^{1/4}$

In exercises 8 – 10, simplify and write the given term without using radicals.

8. $\left(\sqrt{x} + \dfrac{1}{\sqrt{x}}\right)^2$

9. $(\sqrt{x})^2 + \left(\dfrac{1}{\sqrt{x}}\right)^2$

10. $\left(\sqrt[3]{x} + 1\right)^3$

In exercises 11 – 17, simplify the given term and write your answer without negative exponents.

11. $\left(\dfrac{-5x^{-1/4}y^3}{x^{1/4}y^{1/2}}\right)^2$

12. $\left(\dfrac{-2x^{2/3}y^2}{x^{-2}y^{1/2}}\right)^6$

13. $\left(\dfrac{-3s^{2/3}t^2}{4s^3t^{5/3}}\right)^3$

14. $-3(x^2 + 4x + 4)^{-4}(2x + 4)$

15. $\dfrac{1}{3}(x^4)^{-2/3}(4x^3)$

16. $\dfrac{(e^{x+3})^2}{e^{-x}}$

17. $\dfrac{e^{x^2-1}}{e^{x+1}}$

In exercises 18 – 20, simplify and write the given term in exponential form.

18. $\dfrac{4x - 1}{\sqrt[3]{(3x + 2)^2}}$

19. $\sqrt[3]{\left(\dfrac{e^{4\theta-6}y^2}{e^\theta y^{-4}}\right)}$

20. $\sqrt[4]{\left(\dfrac{x^2y^5}{y^{-3}}\right)^2}$

1.5   Logarithms and Exponential Functions

In this section, we will discuss logarithmic functions and exponential functions. The exponent rules we learned last section also apply to the exponents we see in exponential functions, so here we will focus on the relationship between exponential and logarithmic functions. As we mentioned previously, these functions are inverses of each other, in the same sense that square roots and squaring are inverses of each other.

Logarithmic functions and exponential functions are both used to describe many applications such as population growth and value of investments over time. Logarithmic are very prevalent in later courses like differential equations, so a solid understanding of their properties now will help you prepare for these later courses.

The Relationship Between Logarithmic and Exponential Functions

We saw earlier that an exponential function is any function of the form $f(x) = b^x$, where $b > 0$ and $b \neq 1$. A logarithmic function is any function of the form $g(x) = \log_b (x)$, where $b > 0$ and $b \neq 1$. It is no coincidence that both forms have the same restrictions on b because they are inverses of each other. This means that, for the same value of b, $b^{\log_b (x)} = x$ for $x > 0$ and $\log_b b^x = x$.

We also discussed two commonly used logarithmic functions that have special notation. The logarithmic function $f(x) = \log (x)$ is a special way of writing $f(x) = \log_{10} (x)$ and $g(x) = \ln x$ is a special way of writing $g(x) = \log_e (x)$. These also have special names; $\log (x)$ is called the common logarithm and $\ln (x)$ is called the natural logarithm. Note that e is just a number: $e \approx 2.71828$; e is an irrational number, just like π, meaning it can't be written as a fraction of whole numbers.

Let's look at some concrete examples to help illustrate this relationship. Let's look at $b = 2$. For $b = 2$, $f(x) = 2^x$. Then, we can see that $f(3) = 2^3 = 8$. Since they are inverse functions, $g(8) = \log_2 (8) = 3$. In general this is explained as:

$$\text{If } a = b^c, \text{ then } c = \log_b (a). \tag{1.39}$$

newpage

Another way of thinking about it is to use the question "y is b raised to what power?" when you see $\log_b (y)$. For example, when we see $\log_2 (16)$, we ask "16 is 2 raised to what power?" Through a bit of guess and check we get $2^1 = 2$, $2^2 = 4$, $2^3 = 8$, and $2^4 = 16$. This tells us that 2 raised to the 4th power gives us 16, so $\log_2 (16) = 4$. Let's look at a few more examples.

Example 37   Evaluating Logarithmic Functions
Evaluate each of the following statements:

Notes:

1. $\log_3 (9)$

2. $\log_3 \left(\frac{1}{9}\right)$

3. $\log_9 (3)$

4. $\log_2 (2)$

5. $\log_2 \left(\frac{1}{2}\right)$

6. $\log_2 1$

Solution   We'll work on these one at a time.

1. For this statement, we are answering the question "9 is 3 raised to what power?" We can see through a little bit of trial and error that $3^2 = 9$, so we get that

$$\log_3 (9) = 2$$

2. For this statement, we are answering the question "$\frac{1}{9}$ is 3 raised to what power?" We saw in the previous question that $\log_3 (9) = 2$ which gives us a hint that our answer is related to 2, but that we need 3^2 to be in the denominator. Since $3^{-2} = \frac{1}{9}$, we get that

$$\log_3 \left(\frac{1}{9}\right) = -2$$

3. For this statement, we have a different base. Here our base is 9, so we are answering the question "3 is 9 raised to what power?" You may recognize that $\sqrt{9} = 3$; we saw in our last section that another way of writing $\sqrt{9}$ is $9^{1/2}$. This tells us that

$$\log_9 (3) = \frac{1}{2}$$

4. For this statement, we are answering the question "2 is 2 raised to what power?" We can quickly see that $2^1 = 2$, so we get that

$$\log_2 (2) = 1$$

Notes:

5. For this statement, we are answering the question "$\frac{1}{2}$ is 2 raised to what power?" This one is a bit tricky, but we've seen something similar in question 2. In question 2, we saw that because we had a fraction with a power of 3 (the base we were working with) in the denominator, that our final answer was negative. Here if we try 2^{-1}, we get $\frac{1}{2}$. This tells us that

$$\log_2 \left(\frac{1}{2} \right) = -1$$

6. For this statement, we are answering the question "1 is 2 raised to what power?" We know our answer can't be a positive integer because 2 raised to a positive integer gets bigger, and we know it can't be a negative integer since 2 raised to a negative integer gives numbers less than 1. Let's see what happens if we try zero: $2^0 = 1$, so we have that

$$\log_2 (1) = 0$$

We recommend looking through these questions and identifying patterns. When did we get a positive answer? When did we get a negative answer? When did we get a fractional answer? Can you try out some similar problems and see if your answers fit the patterns you identified? We didn't have any answers that were negative fractions; can you come up with such a problem? Notice that all of our questions had positive inputs; it is not possible to find an answer with a negative input. Why? Let's think back to the question we asked for each problem above. If we try to evaluate $\log_2 (-2)$ we would ask "-2 is 2 raised to what power?" Is there an exponent, say x, where $2^x = -2$? No! We can't take a positive number and raise it to a power and end up with a negative number.

Logarithm Rules

We have five main rules that we will need to use when working with logarithms. These rules are all based off of the rules we have for exponents. Be sure to be careful of the details in each of these rules; in some of them all of the logarithms have the same base, but in others the base changes to show the relationship between different logarithmic functions. As we have already seen, the base plays a big role in the specific meaning of the function, so be aware of the rules that have the same base everywhere and the rules where the base changes. Additionally, remember that every base must be positive and not equal to one; also, all inputs must be positive.

Notes:

- $\log_b (x^a) = a \log_b (x)$

- $\log_b (xy) = \log_b (x) + \log_b (y)$

- $\log_b \left(\dfrac{x}{y} \right) = \log_b (x) - \log_b (y)$

- $\log_a (b) = \dfrac{1}{\log_b (a)}$

- $\log_b (x) = \dfrac{\log_c (x)}{\log_c (b)}$

All of these rules can be used in either direction; you can start with the form on the left and rewrite as the form on the right or you can start with the form on the right and rewrite as the form on the left. We won't explain all of these in detail, but we will illustrate examples for the first two.

With the first rule, we can get insight from the question we used earlier to evaluate our logarithms. We'll look at a concrete example rather than a general case. Suppose we want to evaluate $\log_2 (8^4)$. Before we get to our question, let's rewrite this a bit. We can write $\log_2 (8^4) = \log_2 ((2^3)^4)$ since $8 = 2^3$. Next, we can use our exponent rules to get $\log_2 ((2^3)^4) = \log_2 (2^{12})$. Now, answering our question " 2^{12} is 2 raised to what power?," we get 12. So, overall, we have $\log_2 (8^4) = 12$. Now, working on the other side, we have

$$12 = 4 \times 3$$
$$= 4 \log_2 (8)$$

Putting both pieces together, we have $\log_2 (8^4) = 4 \log_2 (8)$.

The second rule also follows from exponent rules. Let's take a look:

$$\log_2 (2^3) + \log_2 (2^4) = 3 + 4 \text{ from the definition of log base 2}$$
$$= 7$$
$$= \log_2 (2^7)$$
$$= \log_2 (2^3 \times 2^4) \text{ from our exponent rules}$$

We illustrated each of these rules using some easy to work with values, but they are true for all values, as long as we have a positive base that is not one and avoid negative inputs. The other rules can all be illustrated in similar ways. (Note: you may want to try to come up with your own examples for these rules to help you understand why these rules are true.) Let's look at an example where we put these rules to use.

Notes:

Example 38   Combining Logarithms

Write $5\log_2(x) + 3\log_2(2y)$ as a single logarithm.

Solution   Currently, this term is the sum of two logarithms, both with the same base, and we want to write it as a single logarithm. It looks like we may want to start with the second rule, $\log_b(xy) = \log_b(x) + \log_b(y)$. It looks like we are already in the form on the right side. However, there is one issue. Currently, both of our logarithms are multiplied by a constant, and the second rule doesn't have coefficients on the logarithms. We'll need to use the first rule to move these coefficients inside of the logarithms before we use the second rule. We get

$$
\begin{aligned}
5\log_2(x) + 3\log_2(2y) &= \log_2(x^5) + \log_2((2y)^3) \\
&= \log_2(x^5) + \log_2(8y^3) \\
&= \log_2[(x^5)(8y^3)] \\
&= \log_2(8x^5y^3)
\end{aligned}
$$

Notice that when the 3 on the second term came inside, we were careful to apply it as an exponent to everything inside of the logarithm, and not just the y. We get that

$$
\boxed{5\log_2(x) + 3\log_2(2y) = \log_2(8x^5y^3)}
$$

Sometimes we will want to go in the opposite direction and split a single logarithm into the sum or different of many logarithms. Let's take a look at an example.

Example 39   Splitting Logarithms

Expand $\ln\left(\dfrac{2x^3y^3}{wz^5}\right)$ into the sum and/or difference of multiple logarithms.

Solution   Here, we want to rewrite as many simpler logarithms. First, we see that the logarithms has a quotient inside, so we can use the third rule to split it:

$$
\ln\left(\frac{2x^3y^3}{wz^5}\right) = \ln(2x^3y^3) - \ln(wz^5)
$$

Notes:

Now, we have products inside of both terms, so we can use the second rule to split these:

$$\ln\left(2x^3y^3\right) - \ln\left(wz^5\right) = \ln\left(2\right) + \ln\left(x^3y^3\right) - \ln\left(wz^5\right)$$
$$= \ln\left(2\right) + \ln\left(x^3\right) + \ln\left(y^3\right) - \ln\left(wz^5\right)$$
$$= \ln\left(2\right) + \ln\left(x^3\right) + \ln\left(y^3\right) - \left[\ln\left(w\right) + \ln\left(z^5\right)\right]$$
$$= \ln\left(2\right) + \ln\left(x^3\right) + \ln\left(y^3\right) - \ln\left(w\right) - \ln\left(z^5\right)$$

Now, for the last step, we can bring the exponents to the outside of each term, giving us

$$= \ln\left(2\right) + 3\ln\left(x\right) + 3\ln\left(y\right) - \ln\left(w\right) - 5\ln\left(z\right)$$

Notice that we were careful to use parentheses around $\ln\left(wz^5\right)$ when we split it because of the negative sign. In the original form we are dividing by w and by z^3 so we need to make sure both of these terms are subtracted when we split the logarithms. Also, we did not evaluate $\ln\left(2\right)$. Since $\ln\left(2\right)$ is really $\log_e\left(2\right)$ and 2 is not made by raising e to an integer or fraction, $\ln\left(2\right)$ is a non-repeating decimal. This means it is better to leave $\ln\left(2\right)$ in our answer than to use a calculator to evaluate it because this form is more precise. This means that our final answer is

$$\boxed{\ln\left(\frac{2x^3y^3}{wz^5}\right) = \ln\left(2\right) + 3\ln\left(x\right) + 3\ln\left(y\right) - \ln\left(w\right) - 5\ln\left(z\right)}$$

Solving Exponential Statements

Logarithms are also used to solve exponential statements, statements where the variable is part of an exponent. When solving an exponential statement, we first need to isolate the exponential term. Once we have isolated the exponential term, we can take a logarithm of both sides. We don't want to take just any logarithm, we want to use a logarithm that has the same base as the exponent so that we can easily simplify our final answer. After we have taken a logarithm of both sides, we can use our logarithm rules to bring the exponent (which has the variable) outside of the logarithm so that we can solve for the variable. Let's take a look.

Example 40   **Solving an Exponential Statement**
Solve $5^{3x-1} - 2 = 0$ for x.

Notes:

Solution   First, we will need to isolate the exponential term, 5^{3x-1}.
Then, we will take log base 5 of both sides since the exponent has 5 as its base.

$$5^{3x-1} - 2 = 0$$
$$5^{3x-1} = 2$$
$$\log_5\left(5^{3x-1}\right) = \log_5(2)$$

Now, we will use our logarithm rules to bring x outside of the logarithm. This
gives

$$(3x - 1)\log_5(5) = \log_5(2)$$
$$(3x - 1)(1) = \log_5(2)$$
$$3x - 1 = \log_5(2)$$
$$3x = \log_5(2) + 1$$
$$x = \frac{\log_5(2) + 1}{3}$$

Notice that when we brought the exponent outside of the logarithm, we
kept the entire exponent inside of parentheses. This is to make sure that we do
not incorrectly distribute terms. Additionally, notice that our final answer still
includes a logarithm term. This is because $\log_5(2)$ does not evaluate to a "nice"
number, so it is more precise to write our final answer this way rather than using
a calculator or computer to evaluate that term. Our final answer is

$$\boxed{5^{3x-1} - 2 = 0 \text{ solves to give } x = \frac{\log_5(2) + 1}{3}}$$

Notice that in this example, we end up with $\log_5(5)$ as part of our work. We
know that this simply evaluates to 1. This is why we used log base 5, and not a
different logarithm. Any logarithm would allow us to solve for x, but using log
base 5 makes it easier to simplify our final answer.

Sometimes you will need to solve for a statement that has two exponential
terms. When this happens, you may be able to employ a useful technique to
solve. Let's take a look at an example.

Example 41   Solving an Exponential Statement
Solve $4^{2y+1} = 2^{y-1}$ for y.

Notes:

Solution   With these types of problems, we want to look at both bases and see if they are related in any way. Here, we have a base of 2 on the right and a base of 4 on the left. You'll probably notice that $4 = 2^2$; we can use this to our advantage when solving. Let's start by rewriting our statement using this fact.

$$4^{2y+1} = 2^{y-1}$$
$$(2^2)^{2y+1} = 2^{y-1}$$
$$2^{2(2y+1)} = 2^{y-1}$$
$$2^{4y+2} = 2^{y-1}$$

Notice that we are using our exponent rules here, specifically the rule $(x^a)^b = x^{ab}$. This allows us to rewrite the statement so that both terms have the same base. Since the two terms have the same base and are equal to each other, we know that they must have equal exponents. This gives us

$$4y + 2 = y - 1$$
$$3y = -3$$
$$y = -1$$

Our final answer is

$$\boxed{4^{2y+1} = 2^{y-1} \text{ solves to give } y = -1}$$

We could solve this problem without using this technique. We would want to take either log base 2 of both side or log base 4 of both sides. Then, we would need to use logarithm rules to simplify and bring the y terms outside of the logarithm before we solve. Both methods result in the same answer; you can practice your logarithm skills by using this alternative method and comparing your final answer to the one above; they should be identical.

Solving Logarithmic Statements

A logarithmic statement is a statement in which the variable of interest is an input to a logarithm. As we know, logarithms and exponential functions are closely related, so it's no surprise that we will use exponential functions to help solve logarithmic statements. Here, we will again use the fact that they are inverse functions, as shown by our definition of a logarithm. Let's look at an example.

Notes:

Example 42   **Solving a Logarithmic Statement**
Solve $\log_5 (2x + 3) = 2$ for x.

Solution   Here, the logarithm is already isolated on one side, so we can start off by using the definition of logarithms shown in equation 1.39 to remove the logarithm from our equation.

$$\log_5 (2x + 3) = 2, \text{ then, from our definition,}$$
$$2x + 3 = 5^2$$
$$2x + 3 = 25$$
$$2x = 22$$
$$x = 11$$

Our final answer is

$$\boxed{\log_5 (2x + 3) = 2 \text{ solves to give } x = 11}$$

Remember when working with either logarithms or exponential functions that they are strongly tied together: when solving a logarithmic statement we need to use exponents and when solving exponential statements we need to use logarithms.

Notes:

Exercises 1.5

Terms and Concepts

1. Explain the relationship between logarithmic functions and exponential functions.

2. What questions are you answering when you evaluate $\log_5(25)$?

3. What is the value of the base for $\ln(x)$?

4. Explain why logarithms help solve exponential statements.

Problems

Evaluate the given statement in exercises 5 – 8.

5. $\log_3(81)$

6. $\ln(e^{5.7})$

7. $e^{-\ln(x)}$

8. $4^{\log_2(2^2)}$

Write the given statement as a single simplified logarithm in exercises 9 – 12.

9. $4\log_3(2x) - \log_3(y^2)$

10. $\dfrac{2}{3}\ln(x) + 3\ln(2y)$

11. $(2x)\log_2(3) + \log_2(5)$

12. $3\ln(xy) - 2\ln(x^2y)$

In exercises 13 – 17, solve the given problem for x, if possible. If a problem cannot be solved, explain why.

13. $5^x = 25$

14. $5^x = -5$

15. $5^x = 0$

16. $5^x = 0.2$

17. $5^x = 1$

In exercises 18 – 25, solve the given problem for x.

18. $3^{x-6} = 2$

19. $4^{2x-5} = 3$

20. $2^{5x+6} = 4$

21. $6^{x+\pi} = 2$

22. $\left(\dfrac{1}{6}\right)^{-3x-2} = 36^{x+1}$

23. $-15 = -8\ln(3x) + 7$

24. $2^x = 3^{x-1}$

25. $8 = 4\ln(2x+5)$

2: Basic Skills for Calculus

In this chapter, we will look at several basic skills and topics that will be used often in calculus: linear functions, solving inequalities, function domains, graphs and graphing, and completing the square. Lines and linear functions appear quite often in calculus. Secant lines are used to determine how fast a function is changing over an interval, tangent lines are used to determine how fast a function is changing at a single point, and linear functions are used to approximate more complicated functions. You will need to solve inequalities to help determine key characteristics of a function, such as when it is increasing and when it is decreasing. Function domains will be useful in working with applied problems to make sure your model has both real world and mathematical meaning. Graphs and graphing will help you identify key features of functions like maximum and minimum values. Completing the square will show up in integral calculus when you need to have your function in a particular form.

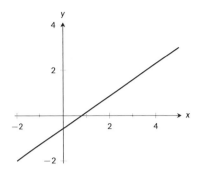

Figure 2.1: A line with a positive slope.

2.1 Linear Functions

In this section, we will discuss linear functions. A *linear function* is a polynomial with degree 1. In calculus, you will learn how to use lines to approximate more complicated functions in order to better understand their behavior at or near a point. These linear approximations can help us evaluate nearby points on the complicated function more easily and can tell us how quickly the function is changing. Linear approximations can also be used to help us determine the area of irregular shapes.

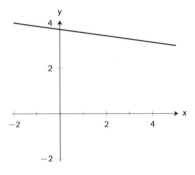

Figure 2.2: A line with a negative slope.

Properties of Lines

Every line can be uniquely defined based on two keep features: the slope of the line and a point contained by the line. The slope of the line tells us about its steepness, and the point gives us a place to anchor the line. If the points on a line go up as you move to the right, it has a positive slope, and the bigger the slope is the faster the line increases, or the steeper it is. You can see an example of a line with positive slope in Figure 2.1. If the points on a line go down as you move to the right, it has a negative slope. A more negative slope indicates that it decreases faster. If the points neither go up nor down as you move to the right, it has a slope of 0, indicating that the line is horizontal.

Visually, it's straightforward to determine if the slope is positive, negative, or 0. Typically, we need more detailed information and will need to calculate the slope. To do this from a graph, we need to find two points on the line. We'll

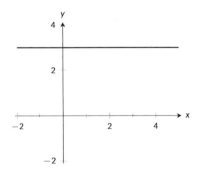

Figure 2.3: A line with a slope of 0.

call these two points (x_1, y_1) and (x_2, y_2). Notice that each point includes two values, an x-value and a y-value. When you pick two points, you may use any two distinct points you like; no matter what you pick you will always get the same result for the slope. Mathematicians typically use m as the symbol for slope. The formula is:

$$m = \frac{y_1 - y_2}{x_1 - x_2} \qquad (2.1)$$

The numerator tells us how much change we have in y (vertical change) between the points and the denominator tells us how much change we have in x (horizontal change). Sometimes, you'll see the formula written like this:

$$m = \frac{\Delta y}{\Delta x} \qquad (2.2)$$

The symbol Δ (the capital Greek letter "delta") tells us we are looking at change, so Δy means change in y and Δx means changes in x. Formulas 2.1 and 2.2 have the same meaning, they just look a little different.

Example 43   Finding Slope
Determine the slope of the line in Figure 2.1.

   Solution   To determiner the slope, we will need two points from the graph. Any two points will work, but we will use $(-2, -2)$ and $(5, 3)$ since both x and y are integer values at these points. We'll call $(-2, 2)$ our first point; this means $x_1 = -2$ and $y_1 = -2$. We'll call $(5, 3)$ our second point; this means $x_2 = 5$ and $y_2 = 3$. Then, we just need to plug into the formula for slope:

$$m = \frac{y_1 - y_2}{x_1 - x_2} = \frac{-2 - 3}{-2 - 5}$$
$$= \frac{-5}{-7}$$
$$= \frac{5}{7}$$

Therefore, the slope of the line in Figure 2.1 is

$$\boxed{m = \frac{5}{7}}$$

   Earlier, we talked about how a line is defined by its slope and a point, but slope itself is determined by two points on the line. This means that we can also

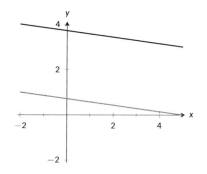

Figure 2.4: Parallel lines.

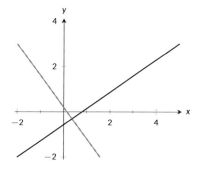

Figure 2.5: Perpendicular lines.

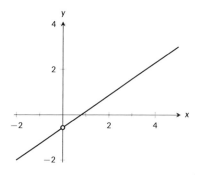

Figure 2.6: The y-intercept of a line.

Notes:

uniquely define a line from two points that are on the line. We can take both to find the slope, and then use either of the points as our anchor.

If two lines have the same slope, we say they are *parallel*. Parallel lines will never intersect each other since they have the same steepness (unless they are really the same line). If two lines are not parallel, they have to intersect at one point. If the lines form a right angle (90°) when they intersect, they are *perpendicular*. Like parallel lines where $m_1 = m_2$ (the slope of line 1 is the same as the slope of line 2), perpendicular lines also have related slopes. Here, if line 1 and line 2 are perpendicular, $m_1 = -\frac{1}{m_2}$. We could rearrange this equation to solve for m_2 and we would get $m_2 = -\frac{1}{m_1}$. Notice that the formula looks the same, except m_1 and m_2 are swapped. This relationship is described as "negative reciprocal;" negative since the sign is opposite and reciprocal since we invert the relationship. We'll see this property used in example 46.

When we look at a line, we can use any two points to describe it, but there are two points that mathematicians are more likely to use. The first (and most commonly used) is the *y-intercept*. This is the point where the line crosses ("intercepts") the y-axis. Since $x = 0$ on the y-axis, this point looks like $(0, y_{int})$. Similarly, the *x-intercept* is frequently used. The x-intercept is the point where the line crosses the x-axis, so it has $y = 0$ and looks like $(x_{int}, 0)$. These two points are commonly used because they are easier to work with since one of the coordinates is 0.

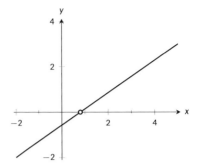

Figure 2.7: The x-intercept of a line.

Expressing Linear Functions

There are two common ways of expressing linear functions, point-slope form and slope-intercept form. As you can probably guess from the names, both of these forms will require that you first know the slope. Slope-intercept form requires knowing the y-intercept of the function, and point-slope form allows you to use any point on the line. It's good to be comfortable working with both forms and with switching between them because sometimes one form will be more useful than the other. In calculus, point-slope form often comes in handy because we will be looking at lines over a small region that won't always contain the y-intercept of the line. Slope-intercept form is:

$$y = mx + b \tag{2.3}$$

where m is the slope of the function and b is the y-coordinate of the y-intercept. Point-slope form is:

$$y - y_1 = m(x - x_1) \tag{2.4}$$

where m is the slope and (x_1, y_1) is any point on the line. Let's take a look at an example of how to work with both of these forms.

Notes:

Example 44   Writing an Equation for a Line

Write the equation for the line that passes through $(2, 4)$ and is parallel to $3x + y = 6$ in slope-intercept form.

Solution   Let's analyze the information we so far. First, we know that our line goes through the point $(2, 4)$. We know we also need its slope.

We're told it's parallel to $3x+y = 6$, so it will have the same slope as that line. However, this line isn't in either of our forms, so it's not immediately clear what the slope is. We'll start by putting this line, $3x + y = 6$ into slope-intercept form (because it requires less algebra than putting it into point-slope form would). To put it into slope-intercept form, we need to isolate y. We'll do that by subtracting 3x from both sides. That gives

$$y = 6 - 3x$$

If we rearrange the right side, we get $y = -3x + 6$. Now, it's in slope-intercept form and we can see that the slope is -3.

Now, we have the slope of our line, $m = -3$, and a point on our line, $(2, 4)$, and our goal is to express the line in slope-intercept form. There's a slight problem with this- we don't know the y-intercept. Luckily, we do have enough information to express the line in point-slope form, so we'll start with that and use algebra to get it into slope-intercept form. In point-slope form, we have:

$$y - 4 = -3(x - 2)$$

We'll use algebra to rewrite this in slope-intercept form (i.e., we'll isolate y):

$$y - 4 = -3x + 6$$
$$y = -3x + 6 + 4$$
$$y = -3x + 10$$

So, in slope-intercept form, the line is $y = -3x + 10$. This also tells us that the y-intercept of this line is $(0, 10)$. Our final answer, in slope-intercept form, is

$$\boxed{y = -3x + 10}$$

This isn't the only way to answer this question, so let's try a second method. The two methods will always give you the same result, so it's entirely a matter of preference when choosing which method to use.

Example 45   Writing an Equation for a Line- A Second Method

Write the equation for the line that passes through $(2, 4)$ and is parallel to $3x + y = 6$ in slope-intercept form.

Notes:

Solution From our previous work on this problem, we know that our line goes through the point $(2, 4)$ and that it has a slope of -3. We also know that slope-intercept form looks like

$$y = mx + b$$

and that this statement holds for every (x, y) pair that are on the line. We know m, but we are missing b. However, we do know that $(2, 4)$ is on the line, so

$$y = -3x + b$$

needs to be true for $x = 2$ and $y = 4$. We'll substitute these values into our equation and solve for b:

$$\begin{aligned} 4 &= -3(2) + b \\ 4 &= -6 + b \\ 10 &= b \end{aligned} \tag{2.5}$$

Now, we know b, so our final answer, in slope-intercept form, is again

$$\boxed{y = -3x + 10}$$

Let's look at one more example, this time working with a perpendicular line.

Example 46 **Perpendicular Line**
In slope-intercept form, write the equation of a line with a y-intercept of 5 that is perpendicular to $y - 4 = 6(x - 2)$.

Solution Let's look at the information we have so far. We are told that the line we are interested in has a y-intercept of 5. We know that $x = 0$ for the y-intercept, so this really means that the y-intercept is the point $(0, 5)$. Next, we know that we are perpendicular to the line $y - 4 = 6(x - 2)$. This line is given in point-slope form and has a slope of 6.

The slope of our line has to be $-\frac{1}{6}$ since it's perpendicular to a line with a slope of 6. Overall, that gives us that the equation of our line, in slope-intercept form, is

$$\boxed{y = -\frac{1}{6}x + 5}$$

Notes:

Exercises 2.1

Terms and Concepts

1. Explain the difference between point-slope form and slope-intercept form.

2. To uniquely determine a line, what information do you need?

3. What is the slope of a horizontal line?

4. A line goes through the point $(0, 6)$. Is this the y-intercept of the line or the x-intercept of the line? Explain.

5. Line 1 has a slope of $m_1 = 2$. If line 2 is parallel to line 1, what is m_2?

6. Line 1 has a slope of $m_1 = -4$. If line 2 is perpendicular to line 1, what is m_2?

Problems

In exercises 7 – 16, write an equation for each line in the indicated form.

7. Write the equation in point-slope form for the line that passes through $(1, 2)$ and is parallel to the line $2x + y = 5$

8. Write the equation of the line in slope-intercept form passing through the points $(1, 2)$ and $(-1, 4)$.

9. Write the equation in point-slope form for the line that passes through $(0, 4)$ and is perpendicular to the line $x - 2y = 6$.

10. Write the equation of the line in slope-intercept form passing through the points $(-1, 0)$ and $(3, 6)$.

11. Write the equation of the line in slope-intercept form passing through the points $(-2, 1)$ and $(2, 7)$.

12. Consider the linear function $f(x) = 2x - 8$. What is the value of the function when $x = 0.1$?

13. Write the equation in slope-intercept form for the line that passes through $(-2, 2)$ and is perpendicular to the line $x + 3y = 8$.

14. Write the equation in point-slope form of the line that passing through the points $(3, 6)$ and $(7, 4)$.

15. Write the equation of the line passing through the points $(-4, 4)$ and $(0, -4)$ in slope-intercept form.

16. Write the equation of the line parallel to $y = 6x + 4$ that has a y-intercept of 2 in point-slope form.

In exercises 17 – 20, answer each question about the properties of the given line(s).

17. Consider the linear function $g(x) = -4x + 5$. What is the slope of the function when $x = 4$?

18. Determine the x-intercept of the line $y = 4x - 8$.

19. Determine the y-intercept of the line $y = 4x - 8$.

20. Which line has a steeper slope: $y = 5x + 10$ or the line passing through the points $(-5, 0)$ and $(0, 11)$?

2.2   Solving Inequalities

In this section, we will look at solving inequalities. You will often work with inequalities in calculus, particularly when you work with derivatives. A derivative is a function that tells you how quickly a related function is changing. A positive derivative tells you the function is increasing and a negative derivative tells you the function is decreasing. This means you will need to be able to identify when the derivative is greater than zero and when it is less than zero.

When solving inequalities, mathematicians express their answers using interval notation, a special way of expressing an interval of numbers. The intervals will tell us when the inequality is a true statement, i.e., they tell us all the input values that make the inequality valid. You will also hear mathematicians use the phrase "the inequality holds for...'; this is another way of saying that these are the inputs that make the inequality true. Let's familiarize ourselves with interval notation before we look at inequalities.

Interval Notation

Before we get to solving inequalities, we'll discuss interval notation. *Interval notation* provides us with a way to describe ranges of numbers concisely. Unlike order of operations, with interval notation parentheses and brackets have different meaning. For example, $[1, 4.5]$ is the range of numbers between 1 and 4.5, including those endpoints. For example, 1, 2, π, and 4.5 are all included in that interval, but -1.2, 85, and 4.5000001 are not. However, if we look at $(1, 4.5)$, 2 and π are still in this interval but 1 and 4.5 are not. Brackets tell us we include the endpoint and parentheses tell us that we don't.

With interval notation, we can mix parentheses and brackets if we need to include one endpoint but not the other. For example, $[1, 4.5)$ contains 1 but not 4.5 and $(1, 4.5]$ contains 4.5 but not 1.

Example 47   Interval Notation
Determine if each of the following numbers is included in the interval $[-5, 27)$.

1. 2

2. π

3. -5

4. -8

5. 27

6. 32

7. -5.000001

8. -4.999999

Solution   For each of these, we need to determine if the number is between the two numbers given in the interval.

Notes:

1. 2 is bigger than -5 and smaller than 27 so it is in the interval.

2. π is bigger than -5 and smaller than 27 so it is in the interval.

3. -5 is one of our endpoints, so we need to see if it has a bracket or a parenthesis on that end. It has a bracket, so it is included in the interval.

4. -8 is smaller than -5, so it is *not* included in the interval.

5. 27 is one of our endpoints, so we need to see if it has a bracket or a parenthesis on that end. It has a parenthesis, so it is *not* included in the interval.

6. 32 is bigger than 27, so it is *not* included in the interval.

7. -5.000001 is smaller than -5, so it is *not* included in the interval.

8. -4.999999 is bigger than -5 and smaller than 27 so it is in the interval.

$$2, \pi, -5, \text{ and } -4.999999 \text{ are in the interval}$$
$$-8, 27, 32, \text{ and } -5.000001 \text{ are not in the interval}$$

We can also use interval notation to express ranges that don't have an upper bound. For example, if we wanted to use interval notation to write the range for all positive numbers, we would write $(0, \infty)$. We know there's no limit to how big a positive number can get, so we use ∞ to indicate that we are just looking at numbers bigger than 0. Similarly, we can write $(-\infty, 0)$ to express the range for all negative numbers. Note that for both of these we use a parenthesis with the infinity symbol and not a bracket since infinity isn't a number.

Additionally, we can use interval notation to express more complicated ranges of numbers. We can combine ranges using $\cup$, the shorthand mathematical way of writing "or". For example, $[1, 3] \cup (4, \infty)$ means the range of values between 1 and 3, including the endpoints, as well as any numbers bigger than 4. So 2, 4.1, and 20 are all in this interval, but -2, 3.5, and 4 are not. We can also use notation to limit ranges using $\cap$, the shorthand mathematical way of writing "and also." For example, if we have two ranges, say $(1, 4]$ and $[2, 8)$, and are only interested in the numbers that are in both ranges, we can write $(1, 4] \cap [2, 8)$. We can use this symbol to help show our work, but for our final answer we should always simplify so that we don't need to use the $\cap$ symbol (it's fine, and quite common, to use the $\cup$ symbol as part of your final answer). We said that $\cap$ means we only want the numbers that are in both intervals; the smallest number contained by both intervals is 2 and the largest is 4, so we can write $(1, 4] \cap [2, 8) = [2, 4]$ instead.

Notes:

Interval Notation and Inequalities

Interval notation also gives us another way of expressing an inequality. For example, $x \geq 2$ can be written as $x \in [2, \infty)$. Here the $\in$ symbol is read as the word "in". We would read this out loud by saying "x is greater than or equal to 2" is the same as "x is in the range from 2, inclusive, to infinity." The statement $x \in (2, \infty)$ is a bit different; it's the same as $x > 2$ since we don't want to include 2 as part of our range. Here, we would read the range as "x in 2, exclusive, to infinity." The symbols we learned earlier, $\cup$ and $\cap$ are read as "union" and "intersect," respectively.

When working with inequalities, we will start all inequality problems by turning them into equality problems. This will allow us to use some techniques we've already seen when we discussed factoring and roots of a function. The solution(s) to the equality problem will tell us the "break points," the input values where the inequality may switch from being true to being false. We'll test values on both sides of each break point to see where the inequality is true. We will work carefully to make sure we find all the break points because we don't want to miss any places where the inequality could switch from true to false. Let's look at a few straightforward examples before we move onto the more complicated inequalities.

Example 48   **Polynomial Inequality**

Solve $x^2 - 6x + 8 > 0$.

Solution   Our first step is to convert this into an equality statement by changing the $>$ symbol to an $=$ symbol:

$$x^2 - 6x + 8 = 0$$

Now, we can use any solution method we learned for finding the roots of a quadratic function to solve. Here we have a quadratic that factors nicely, so we will take that approach, but you could use the quadratic formula if you prefer.

$$x^2 - 6x + 8 = 0$$
$$(x - 2)(x - 4) = 0$$
$$x = 2, 4$$

This tells us that the break points are $x = 2$ and $x = 4$. These are the only places that the inequality could change from being true to being false for this type of function. We'll test values on each side of both break points; this means we need to test a value that is less than 2, a value between 2 and 4, and a value bigger than 4. We like to work from left to right, so we will start with testing a

---

Notes:

value less than 2. We can pick any number that is less than 2, but we will use 0 because it is easy to work with. If we substitute in $x = 0$ we get:

$$x^2 - 6x + 8 = (0)^2 - 6(0) + 8 = 8 > 0$$

We get 8, which is bigger than 0, so the inequality is true for all values less than 2. Next, we need to test a value between 2 and 4; 3 seems like the easiest option.

$$x^2 - 6x + 8 = (3)^2 - 6(3) + 8 = 9 - 18 + 8 = -1 < 0$$

We get a negative number, so the inequality is false for everything between 2 and 4. Now, we need to test something bigger than 4. We'll use 5, but you can pick any number, as long as it's bigger than 4.

$$x^2 - 6x + 8 = (5)^2 - 6(5) + 8 = 25 - 30 + 8 = 3 > 0$$

The result is positive, so the inequality is true. Now, we have that the inequality is true for numbers less than 2 and numbers greater than 4. It is not true for $x = 2$ or $x = 4$ since both of these make the left side 0 and we want the left side to be bigger than 0, not equal to it. In interval notation, we have:

$$x \in (-\infty, 2) \cup (4, \infty)$$

In this example, we have a *strict inequality*. We say it is strict because it is $>$ and not $\geq$. Similarly, we would say an inequality with $<$ is strict. With strict inequalities, our final answer will not include the break points, so we will use parentheses at these break points because we do not want to include these points.

Many people will use a number line when working with inequalities. When using a number line, you would mark each break point, and then shade or otherwise mark the intervals where the inequality is true. For the previous problem, this would look like the following:

Since we have a strict inequality (meaning we have $>$ or $<$ so that we are strictly greater than or strictly less than), we use open circles to mark our break points. This reminds us that we do not include these points in our intervals. Some people will use check-marks and X's instead, with a check-marks indicating where the inequality holds and a X where it doesn't. This would look like the following:

Notes:

These number lines become quite useful if you have a lot of break points. They make it very clear so that you can be sure to test a point in each interval. We can also use a table to summarize results, rather than using a number line. The table has a few key advantages: it clearly summarizes your work making your thought process easier to follow and will help eliminate careless errors from your work. A table for the previous example might look like:

	$(-\infty, 2)$	$(2, 4)$	$(4, \infty)$
Value to check:	0	3	5
Result:	$8 > 0$	$-1 > 0$	$3 > 0$
T/F:	True	False	True

Any of these methods are appropriate for clearly showing your work; the one you choose is a matter of personal preference.

Incorporating Undefined Points

We noted earlier that our inequality can change from true to false at our break points, the points where the equality statement is true. The inequality can also change from true to false at places where the function is undefined. For example, we know that the function $f(x) = \frac{1}{x}$ is positive when x is positive and negative when x is negative. This means that the inequality $\frac{1}{x} > 0$ holds, or is true, only for $x \in (0, \infty)$. However, there are no places where $f(x) = 0$. Since $f(x)$ is undefined at $x = 0$, it introduces a different type of break point; one where the graph of the function "breaks" because it cannot be graphed where it is undefined. Let's take a look at an example where we have to incorporate undefined points by including additional break points.

Example 49   **Solving a Rational Inequality**

Solve $\dfrac{x - 5}{x^2 - 4} \geq 0.$

Solution   We'll start by turning the inequality into an equality statement and solving for x. To help solve for x, we will get rid of the fraction by multiplying both sides by the full denominator; this will allow us to cancel the denominator on the left side.

Notes:

$$\frac{x-5}{x^2-4} = 0$$

$$(x^2-4)\left(\frac{x-5}{x^2-4}\right) = (x^2-4)(0)$$

$$x-5 = 0$$

$$x = 5$$

This gives us a break point at $x = 5$. Next, we will need to see if the function is undefined at any points. Since it is a rational function (a fraction with a polynomial in the numerator and a polynomial in the denominator), we know it is undefined anywhere that the denominator equals zero. Let's look for these points:

$$x^2 - 4 = 0$$

$$(x-2)(x+2) = 0$$

$$x = 2, -2$$

We see that $\frac{x-5}{x^2-4}$ is undefined for $x = 2$ and $x = -2$. This gives us two additional break points. That means we have three break points: $x = 5$, $x = 2$, and $x = -2$. Let's mark these on a number line. Since this is not a strict inequality, we will use closed circles to mark the break point at $x = 5$. However, we still need to use open circles at $x = 2$ and $x = -2$ because the function is undefined at these points and we will not include them in our intervals.

Now, we need to check values in each interval. First, we'll check something less than -2; we'll use $x = -3$. Substituting, gives $\frac{(-3)-5}{(-3)^2-4} = \frac{-8}{5}$. This is less than 0; this means we can place an x-mark over this interval:

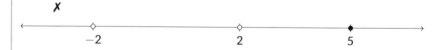

Now, to check something between -2 and 2. We'll use $x = 0$. Substituting gives $\frac{(0)-5}{(0)^2-4} = \frac{-5}{-4} = \frac{5}{4} > 0$. This means we can place a check-mark over this interval:

Notes:

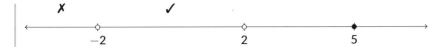

Now, we need to check a value between 2 and 5. We'll use $x = 3$. We get $\frac{(3)-5}{(3)^2-4} = \frac{-2}{5} < 0$, so this interval gets an x-mark.

Lastly, we need to check a value greater than 5. We'll use $x = 6$. This gives $\frac{(6)-5}{(6)^2-4} = \frac{1}{32} > 0$, so this interval gets a check-mark.

We now have a mark over every interval, so we can determine our final answer. We can see that the inequality is true for $x \in (-2, 2) \cup [5, \infty)$. Note that we included $x = 5$ since it has a closed circle and excluded $x = -2$ and $x = 2$ since they have open circles. Our final answer is

$$\frac{x-5}{x^2-4} \geq 0 \text{ holds for } x \in (-2, 2) \cup [5, \infty)$$

Notes:

Exercises 2.2

Terms and Concepts

1. In your own words, explain the what is meant by a strict inequality.

2. In your own words, describe the two ways we can have break points.

3. Does a statement always switch from true to false at a break point? Give an example to support your argument.

4. What methods can you use to find the break points of a quadratic equality?

Problems

In exercises 5 – 11, write each statement in simplified interval notation.

5. $-3 \leq x \leq 10$

6. $x \geq -5$ and $x > 2$

7. $x \geq -5$ and $x < 2$

8. $x \leq -5$ and $x > 2$

9. $x \geq -5$ or $x > 2$

10. $x \leq 4$ and $x > -6$

11. $x > 4$ or $-2 > x$

In exercises 12 – 14, write each statement using inequalities.

12. $x \in [3, 4) \cup (4, \infty)$

13. $x \in [-2, 4)$

14. $x \in (5, 6] \cup [7, 8)$

In exercises 15 – 26, solve the given inequality and express your answer in interval notation.

15. $\dfrac{x - 2}{x - 4} \leq 0$

16. $x^2 - 2x + 8 \leq 2x + 5$

17. $x^2 + 2x > 15$

18. $-x^2 + 7x + 10 \geq 0$

19. $\dfrac{x + 3}{x - 2} - 2 \leq 0$

20. $2x^2 - 4x - 45 \leq -4x + 5$

21. $\dfrac{3x + 1}{x - 2} \leq 2$

22. $1 + x < 7x + 5$

23. $\theta^2 - 5\theta \leq -6$

24. $y^3 + 3y^2 > 4y$

25. $x^3 - x^2 \leq 0$

26. $\dfrac{x^2 + 3x + 2}{x^2 - 16} \geq 0$

2.3   Function Domains

This section covers function domains. In calculus, we will use domains to help identify any discontinuities in functions and perform a full analysis of a function. Function domains will also help identify vertical asymptotes, places where a function may switch between increasing and decreasing, and places where the concavity (general curvature) of a function may change.

Function Domains

The *domain* of a function is the set of all possible real number inputs that result in a real number output for that function. Domains are typically expressed using interval notation, labeled with "*D*:". With domains, it's often easier to look for inputs that will cause problems, rather than looking for "good" inputs. By making a list of trouble points, we will determine the domain by looking at what's left. We'll start by looking at the domain for each of our common functions discussed in Section 1.2.

Domains of Power Functions

For power functions, the domain will depend on the value of the exponent. In Section 1.2, we said that power functions have the form $f(x) = ax^b$ where a and b can be any real numbers. We'll start by looking at where we could run into trouble with certain inputs.

The first place we can run into trouble is if b is negative. This will always cause a problem for $x = 0$ because the negative exponent means we would be dividing by 0 (and we can't do that!).

The second place where we can run into trouble is when b is not a whole number. Here we'll focus on rational numbers, i.e., any number that can be written as a fraction. Say $b = \frac{p}{q}$ where p and q are whole numbers with no common factors. If q is odd, we won't have any trouble with any inputs, but if q is even we can have problems. If we try to take an even root of a negative number (like $\sqrt{-4} = (-4)^{1/2}$) we don't get a real number. We'll have this trouble with any negative input value if q is even, so in this situation, we can't input any negative numbers.

By combining these two trouble spots, we can find the domain of any power function you're likely to run into:

Example 50   Power Function Domains

Determine the domain for each of the following power functions:

Notes:

1. $f(x) = x^{2/3}$

2. $g(x) = 4x^{-2}$

3. $h(x) = -5x^8$

5. $w(t) = \dfrac{1}{2}t^{-1/3}$

6. $y(t) = t^{-3/4}$

7. $z(t) = -2t^{3/4}$

Solution   For each of these, we need to look at the exponent only; scalar multiplication of a function does not affect the domain.

1. For $f(x)$, $b = \frac{2}{3}$. This is a fraction, so we need to look at the denominator. The denominator is 3, an odd number. This tells us that negative inputs are fine. Since b is positive, we know that 0 is also fine. So, we have

$$\boxed{\text{D: } (-\infty, \infty)}$$

2. For $g(x)$, we have $b = -2$. This is a whole number, so we only need to look at its sign. It's negative, so this tells us that 0 will cause trouble. So,

$$\boxed{\text{D: } (-\infty, 0) \cup (0, \infty)}$$

3. For $h(x)$, $b = 8$. This is a positive whole number, so we don't have any trouble inputs since we can only run into trouble if b is negative or a fraction. So,

$$\boxed{\text{D: } (-\infty, \infty)}$$

4. For $w(t)$, we have $b = -\frac{1}{3}$. It's negative, so 0 is a problem, but it's a fraction with an odd denominator, so negative inputs are fine. Thus,

$$\boxed{\text{D: } (-\infty, 0) \cup (0, \infty)}$$

5. For $y(t)$, $b = -\frac{3}{4}$. This is negative, so 0 is a problem. It's a fraction with an even denominator, so negative inputs are also a problem. That leaves

$$\boxed{\text{D: } (0, \infty)}$$

Notes:

6. For $z(t)$, $b = \frac{3}{4}$. This is positive, so 0 is fine, but again it's a fraction with an even denominator so negative inputs are a problem. That means

$$D: [0, \infty)$$

Notice that the domain for $z(t)$ has a bracket, so it includes 0, but the domain for $y(t)$ has a parenthesis so it doesn't include 0.

Domains of Exponential Functions

Next on our list of common functions are exponential functions, functions of the form $f(x) = b^x$, with $b > 0$ and $b \neq 1$.. For exponential functions, we can use any real number input and get a real number as output, so the domain is always $(-\infty, \infty)$.

Domains of Logarithmic Functions

For logarithmic functions, only positive inputs give us real number outputs, so the domain of $\log_b (x)$ is $(0, \infty)$ for every valid base, b. Note that 0 is not in the domain.

Domains of Trigonometric Functions

The last type of common function we discussed was trigonometric functions. Here the domain depends on the exact function you are using; we'll discuss these more later in this text.

Domains for Combined Function

When we look at combined functions, we will start by looking at the domain for each individual function. If an input is a problem for one of the individual functions, it will also be a problem for the combined function. Additionally, other problems can be introduced depending on how the functions are combined. Scalar multiplication, addition, subtraction, and multiplication do not introduce other problems, but quotients and compositions can.

With quotients, for every input we are evaluating a fraction. We can run into a new problem if the denominator is 0. So, as part of determining the domain of a quotient, we will need to see when, if anywhere, the denominator is 0. Let's take a look:

Notes:

Example 51   Quotient Function Domains

Determine the domain of

$$f(\theta) = \frac{\theta^2 + 4}{\sqrt{\theta} - 1}$$

Solution   First, let's look at each of the individual functions. The numerator has $\theta^2 + 4$. This is the addition of two monomials: θ^2 and 4. The addition doesn't introduce any problems. θ^2 is a power function where b is a positive whole number, so it doesn't introduce any problems. 4 doesn't depend on an input, so it doesn't introduce any problems.

The denominator is $\sqrt{\theta} - 1$. This is the difference of $\sqrt{\theta}$ and 1. Like with addition, the difference doesn't introduce any problems. The function 1 doesn't introduce any problems. $\sqrt{\theta}$ is another way of writing $\theta^{1/2}$. This is a power function where b is positive, so 0 is fine. However, b is a fraction with an even denominator so negative inputs cause a problem.

So far, the only issue we have comes from the square root. However, since we have a quotient, we need to see if the denominator is ever 0. We'll do this by solving:

$$\sqrt{\theta} - 1 = 0$$

Adding 1 to both sides gives

$$\sqrt{\theta} = 1$$

Squaring both sides gives us

$$\theta = 1^2 = 1$$

This tells us that we will also have a problem when $\theta = 1$. All together then, we have problems with negative inputs and with 1, so

> The domain of $f(\theta)$ is $\theta \in [0, 1) \cup (1, \infty)$

Notice that in Example 51, 0 is part of the domain even though we have a quotient. A quotient doesn't mean that 0 as an input is a problem, rather that any inputs that make the denominator 0 are problems.

Composition of functions can drastically change domains. With composition, you'll have to restrict the output of the inside function to make sure it's suitable to be an input of the outside function. This can give extra restrictions on the overall domain.

Notes:

Example 52   **Domain of a Function Composition**
Determine the domain of

$$f(x) = \sqrt{6-x} + 12x$$

Solution   Overall, we have the addition of two functions, $\sqrt{6-x}$ and $12x$. Addition doesn't introduce any problems. The second function, $12x$ has no problem inputs because it's a power function where b is a positive whole number. However, we know that the square root function can't use negative inputs since it's really a power functions with $b = \frac{1}{2}$. Since the square root is a composition with $6-x$ as the inside function, we'll need to determine when $6-x$ is negative to find which values of x are a problem. To do that, we'll solve the inequality

$$6 - x < 0$$

Luckily, this isn't too complicated; we'll add $-x$ to both sides to get $6 < x$, or $x > 6$. This tells us that any input bigger than 6 is going to be a problem for $f(x)$. So, the domain of $f(x)$ is $(-\infty, 6]$. We use a bracket on the right since we can use 6 as an input. So, altogether, we have that

> The domain of $f(x) = \sqrt{6-x} + 12x$ is $x \in (-\infty, 6]$

The hardest part of finding domains is working carefully and methodically. You probably noticed that all of these examples seem to have an awful lot of written explanation for math problems. While we don't typically write out complete sentences in our own work, we will include notes like "$\ln(x)$: problems with 0 and $x < 0$". We also keep a running list of problem points on the side of the page as we work through more complicated functions to make sure we get all of the problem inputs so that we can exclude them from the domain.

Notes:

Exercises 2.3

Terms and Concepts

1. What does it mean if $x = 2$ is in the domain of $f(x)$?

2. What does it mean if $x = 4$ is not in the domain of $f(x)$?

3. T/F: The domain of $f(g(x))$ depends only on the domain of $g(x)$. Explain.

4. T/F: The domain of $\frac{f(x)}{g(x)}$ depends only on where $g(x) = 0$. Explain.

Problems

In exercises 5 – 16, express the domain of the given function using interval notation.

5. $\dfrac{\sqrt{3-x}}{x+9}$

6. $\dfrac{\sqrt{x+11}}{x-11}$

7. $\dfrac{\ln(x-6)}{2x-26}$

8. $\dfrac{2t}{\sqrt{t-5}}$

9. $\ln\left(\sqrt{x+3}\right)$

10. $\theta^3 + 4\theta^2 - 20 + \pi$

11. $\dfrac{\log_3(x-4)}{\log_3(2x)}$

12. $\dfrac{x}{\log_2(2x-1)}$

13. $f(x) = \ln(4 - x^2)$

14. $f(x) = \ln(x^2 - 4)$

15. $f(x) = \sqrt{(x+3)^2 - 4}$

16. $f(x) = \sqrt[3]{(x-2)^3 + 1}$

2.4   Graphs and Graphing

In calculus, we will be analyzing graphs to learn more about the functions they represent. It is important that we have a good understanding of the relationship between a function and its graph. Specifically in differential calculus, we will learn to use derivatives (which is a rate of change or the local slope) to determine where a graph is increasing or decreasing. Also, in integral calculus, you might learn how to calculate the work required to pump a fluid, which gives a chance to explore the value of using different locations for the origin, effectively resulting in a shift of the function and its graph.

Graphs of General Functions

Many people take a very tedious approach to graphing; for the domain they are interested in graphing, they take each possible integer value of x, evaluate the function for that value, and then graph a single point. After they have graphed several points across the domain, they will "connect the dots." While this method is reliable, it is time consuming, and can be difficult depending on the function. In this section, we will give an overview of the general shape of common functions and then talk about how these general functions can be shifted, stretched, and flipped in order to quickly sketch related functions. Additionally, we will talk about piecewise functions and how to graph them correctly.

Lines

The quickest way to graph a line is by using a point and the slope. This is largely because both forms, slope-intercept and point-slope, provide you with this information. Start by plotting the point. Then, from that point use the slope to plot a second point. For example, if the slope is $-\frac{2}{3}$, you would start at the initial point, move 3 units to the right (because the horizontal change is 3), and then move down 2 units in the y-direction (because the slope is negative and the vertical change is 2). (Note: you can also move vertically and then horizontally, either way gives the same result.) Plot this new point, and then use a straight-edge to connect both points. Be sure to continue past each of the points. If the slope is a whole number, move that many units vertically and only one unit to the right to plot your second point.

Example 53   **Graphing a Line**
Graph the line $y = 2x - 3$

Solution   Let's start by identifying the slope and a point. We are in slope-intercept form, so we can see that the slope is $m = 2$ and the y-intercept

Notes:

is $(0, -3)$. We'll starting by plotting a point at $(0, -3)$. Then, we'll move to the right 1 unit and up 2 units and plot a second point. Then, we use the two points to draw our line. The full process is shown in the following three graphs:

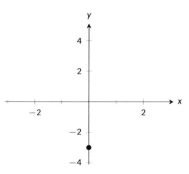

Plot the y-intercept

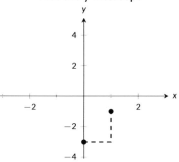

Use the slope to plot a second point

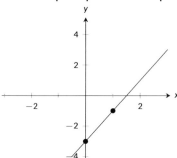

Complete the line

Notes:

Quadratic Functions

The most basic quadratic function is $f(x) = x^2$. Later, we will discuss how every other quadratic function can be graphed by shifting and stretching this function. The general shape of this function is a "U" and it is symmetric over the y-axis, meaning that the left and right sides of the graph are a reflection of each other. This function grows quickly; this means that as x gets big, $f(x)$ gets big faster than x does. It has no horizontal asymptotes, meaning that as x gets big, $f(x)$ doesn't level off.

Cubic Functions

The most basic cubic function is $f(x) = x^3$. Similarly to the quadratic functions, every other cubic function can be graphed by shifting and stretching this function. This function has rotational symmetry around the origin; if you treat $(0, 0)$ like a pivot point and rotate the graph $180°$, it will looks exactly the same. For positive values of x, this function grows quickly, but for negative values of x it becomes more and more negative. It also has no horizontal asymptotes.

Even Polynomials

An *even polynomial* is any polynomial where every monomial has an even degree. For example, $f(x) = x^6 + 3x^4 - 5x^2 + 7$ is an even polynomial because the degrees are 6, 4, 2, and 0. However, $g(x) = x^4 - x^2 + x - 4$ is not an even polynomial because it has x as a term (and x has degree 1). Even polynomials share features with the basic quadratic function: they are symmetric about the y-axis and both "tails" of the function have the same sign. This means that as x gets very big or very negative, $f(x)$ will have the same sign; either both tails are positive or both tails are negative. This is a key feature of even polynomials. The sign on the highest degree term tells you if both of the tails will be positive or if both will be negative.

A related, but slightly different, type of function is a *polynomial of even order*, also known as a polynomial of even degree. Here, we only care about the highest degree term being even, so both $f(x)$ and $g(x)$ from above are polynomials of even order. These polynomials don't have to be symmetric about the y-axis, but they do exhibit the same "end" behavior where either both tails are positive or both tails are negative.

Odd Polynomials

An *odd polynomial* is any polynomial where every monomial has an odd degree. For example, $f(x) = 4x^5 + 2x^3 - 7x$ is an odd polynomial because the

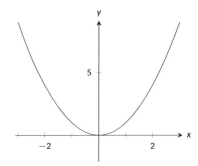

Figure 2.8: The graph of $f(x) = x^2$.

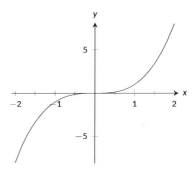

Figure 2.9: The graph of $f(x) = x^3$.

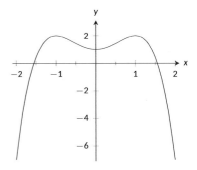

Figure 2.10: The graph of $f(x) = -x^4 + 2x^2 + 1$, an even polynomial.

Notes:

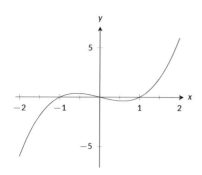

Figure 2.11: The graph of $f(x) = x^3 - x$, an odd polynomial.

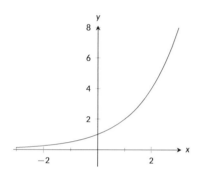

Figure 2.12: The graph of $f(x) = 2^x$, a basic exponential function.

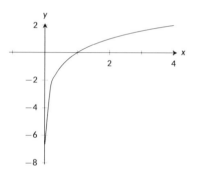

Figure 2.13: The graph of $f(x) = \log_2(x)$, a basic logarithmic function.

degrees are 5, 3, and 1. However, $g(x) = 2x^3 + x - 7$ is not and odd polynomial because it has -7 as a term (and -7 has degree 0). Odd polynomials share features with the basic cubic functions: they have rotational symmetry about the origin and the tails have opposite signs. Odd polynomials always have one positive tail and one negative tail.

Again, we have a related type of function, a *polynomial of odd order*, also known as a polynomial of odd degree. For these polynomials, the highest degree must be odd, but smaller degree terms can be even, as in $h(x) = x^3 + 2x^2$. These polynomials aren't all symmetric about the origin, but the tails will have opposite signs.

Exponential Functions

For a basic exponential function, $f(x) = b^x$, b must be a positive real number with $b \neq 1$. All basic exponential functions with $b > 1$ share some key features: they all contain the point $(0, 1)$, they all contain the point $(1, b)$, they grow quickly for positive values of x, and for negative values of x they get close to $y = 0$, the x-axis. A basic exponential function will never cross the x-axis, it will only get closer and closer as x gets more and more negative. This long-term behavior is described as having an *horizontal asymptote* at $y = 0$. When graphing an exponential function, we plot the two key points listed above and use the general shape to guide the rest of our graph.

Logarithmic Functions

Logarithmic functions have the same shape as basic exponential functions, but reflected over the line $y = x$. This means that the key features of functions of the form $f(x) = \log_b(x)$ are: they all contain the point $(1, 0)$; they all contain the point $(b, 1)$; for very small positive values of x, $f(x)$ becomes increasingly negative if $b > 1$ and increasingly positive if $b < 1$; as x becomes very large, so does $f(x)$ if $b > 1$ and very negative if $b < 1$. It's important when graphing logarithmic functions to remember that their domain is only $(0, \infty)$; you should graph nothing for negative values of x and nothing for $x = 0$. Logarithmic functions have a *vertical asymptote* at $x = 0$, the y-axis. The graph gets very close to this vertical line, but it will never cross it.

Trigonometric Functions

A key feature of the trigonometric functions $\sin(x)$, $\cos(x)$, and $\tan(x)$ is that all three are *periodic* functions; they exhibit the same pattern over and over. Additionally, the sine and cosine functions never grow without bound; their y values are always between -1 and 1. The tangent function does grow without

Notes:

bound and has repeat vertical asymptotes. For all three functions, there are key points when x is a multiple of π, such as $x = \frac{\pi}{2}$, $x = \pi$, $x = \frac{3\pi}{2}$, and $x = 2\pi$. Additionally, you might notice that the graphs of sine and cosine are very similar. In fact, if you take the graph of sine and shift it to the left by $\frac{\pi}{2}$ you would get the graph of cosine. In fact, this is a trigonometric identity that can be seen just from the graphs. The graphs of these three functions can be seen below:

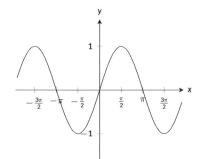

The graph of $f(x) = \sin(x)$

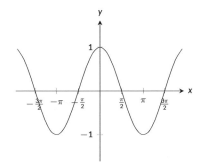

The graph of $f(x) = \cos(x)$

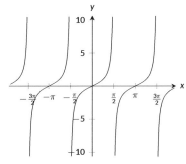

The graph of $f(x) = \tan(x)$

Modifying Functions

Now that we've seen the shapes of some basic functions like $f(x) = x^2$, let's talk about how we can use these basic shapes to help graph many functions. We'll need to be able to identify the base function we are working with and to identify how it's been modified. We'll break these down into two main types of modifications: vertical modifications and horizontal modifications. Vertical modifications will, as the name says, will affect the functions vertically, either shifting the function up or down, or stretching or shrinking the function's height. Similarly, horizontal modifications will affect the function horizontally, shifting it left or right, or stretching or shrinking its "width."

Vertical Modifications

The first type of vertical modifications we will discuss are shifts, where every point on the functions gets shifted up or down by the same distance. This is one of the easiest modifications to spot; all we have to do is add or subtract a constant to the function. Let's take a look at an example.

Example 54   Shifting a Function Vertically
Graph the function $f(x) = x^2 + 3$.

Notes:

Solution   Here we can see that we have a function with a constant added to it; this tells us that we need to graph x^2, but with a vertical shift. The constant, $+3$ tells us that we will take this base function and shift it up 3 units (if this was -3 we would shift the function down by 3 units).

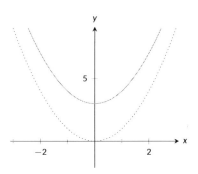

Here, the graph shows both our base function, x^2 (dotted) and our desired, shifted function, $f(x) = x^2 + 3$ (solid). If you take a piece of wire and shape it to match the graph of x^2, you can move the wire up 3 units and see that you get exactly the graph of $f(x) = x^2 + 3$.

Notice that with a vertical shift, the shape of the function does not change at all, only its positioning relative to the x axis changes. With our next modification, vertical stretching and shrinking, we don't change its positioning, or the shape, but we do change how steep the function is. With vertical stretches, everything on the x-axis is fixed, so the positioning doesn't change. The shape doesn't change in the sense that every quadratic will still look like a "U" and every sine or cosine will still look like never-ending waves (similarly for our other functions). So how do we recognize when a function is being stretched or shrunk vertically? We'll have a base function that is being modified with scalar multiplication, such as $f(x) = 3\sin(x)$ or $g(t) = -\frac{1}{2}t^3$.Here, if we multiply by a number bigger than 1 or less than -1, we will stretch the function and make it steeper. If we multiply it by anything between -1 and 1, it will shrink and get less steep. If we multiply by a negative number, not only is the function being stretched or shrunk, it will also be flipped; everything above the x-axis will be reflected to below the x-axis and everything below will be reflection to above the x-axis.

Example 55   **Stretching/Shrinking a Function Vertically**
Graph the functions $f(x) = \frac{1}{2}x^3$ and $g(x) = -\frac{1}{2}x^3$.

Solution   For both $f(x)$ and $g(x)$ we have the same base function, x^3. Since $f(x)$ is formed by multiplying this by $\frac{1}{2}$, it is being shrunk, but not flipped.

Notes:

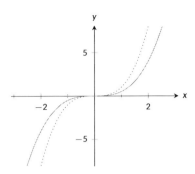

Here we can see our base function, x^3 (dotted) and $f(x) = \frac{1}{2}x^3$ (solid). You'll see that the shape and position are still the same, but $f(x)$ stays closer to the x-axis; it does not get tall as quickly as x^3 does because we shrunk the graph vertically. Now that we've shrunk the graph, flipping it will give us the graph of $g(x)$ (solid):

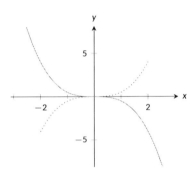

This graph has both $f(x)$ (dotted) and $g(x)$ (solid); both have the same position and the same steepness, but $g(x)$ is upside-down.

The previous example shows how to shrink and flip a graph. Here, you could do either step first; if you flip and then shrink you would get the exact same result. We recommend making very quick sketches in the margins of your page when dealing with multiple transformations at the same time; it makes it easier for me to make sure we draw the final graph accurately by capturing each stage, however, after practice you may feel comfortable doing both steps at once.

We've now seen how we vertical modifications work individually, but what happens when we combine them?

Example 56   **Multiple Vertical Modifications**
Graph $h(x) = 2\sin(x) - 1$.

Notes:

Solution   First, let's identify our base function. Here, we are working with sin (*x*). We see that we are multiplying by 2 and subtracting 1; this tells us we have a vertical stretch and a vertical shift. Which should we do first? The answer comes from our order of operations: multiplication should be done before subtraction. We'll follow that same rule here by stretching sin (*x*) and then shifting it. Since we have 2 sin (*x*) in our function, we will start by graphing sin (*x*) (dotted) and then making it twice as "tall" (i.e., twice as far from the x-axis; solid graph).

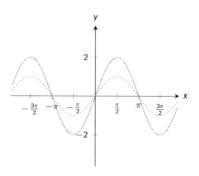

Now that we've stretched it, we can take care of the addition and shift it down 1 unit:

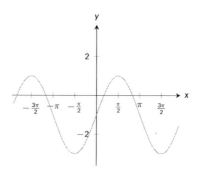

Horizontal Modifications

We can make the same modifications horizontally that we made vertically: stretching, shrinking, and shifting. With vertical modifications, the modifications showed up on the outside of the function, with shifts added to the end and with stretches coming as multiplication out front. With horizontal changes we will

Notes:

work on the inside of the function. For example, if we want to shift the function $f(x)$ 2 units to the left, we would graph $f(x+2)$. This moves the graph of f to the left because, in essence, we always looking at a bigger input than what x really is, e.g., if $x = 4$, really we are looking at $f(4+2) = f(6)$. Similarly, if we want to shift f to the right, we would use a subtraction: $f(x-2)$.

Similarly, if we want to stretch or shrink the function horizontally, the change will also show up on the inside. To stretch a function by $a > 0$, we would graph $f(\frac{1}{a}x)$. We use $\frac{1}{a}$ because in order to stretch it horizontally we need x to change more slowly. In order to shrink it by a factor of $a > 0$ we would multiply: $f(ax)$. If we want to flip the graph horizontally, we will still multiply by a negative: $f(-x)$. Notice that many of the horizontal modifications don't immediately work the way you would expect, unlike the vertical modifications. If you are feeling a bit confused by these, we would recommend graphing a few using the point by point method. Similarly,the process for dealing with multiple modifications is a bit different than what you might expect: first we will identify the base function, then include any shifts, and then include any stretches or shrinking. Let's look at an example:

Example 57   Multiple Horizontal Modifications
Graph the function $f(x) = 4x^2 + 4x + 1 = (2x+1)^2$.

Solution   Our base function here is x^2 (dotted). We don't have any vertical modifications, just horizontal modifications since everything is happening inside the function. We see we have a shift left of 1, due to the $+1$, and then we need to shrink by a factor of 2 since x is multiplied by 2. First, we include the shift (solid):

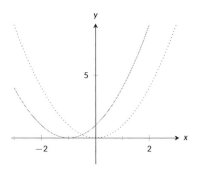

then, we shrink the function horizontally (solid):

Notes:

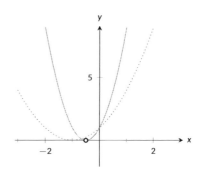

Notice that when we shrink it, the point on the y-axis, $(0, 1)$, is the only point that is the same between both graphs. This is because we shrink and stretch around the y-axis, not around the center of the graph. We can see this also by seeing that the x-intercept changes from $(-1, 0)$ to $(-\frac{1}{2}, 0)$ (labeled with an open dot).

If a function is being modified both vertically and horizontally, you should take care of all the horizontal changes first. Let's see how this looks.

Example 58   **Graph Transformation**
Graph the function $g(x) = -(\frac{1}{2}x + 2)^2 + 3$.

Solution   As with our previous examples, the first step is to identify the base function. Here our base functions is x^2 (dotted). We said that we should start with horizontal changes, so let's look at those first. With horizontal modifications, we need to work with the shift and then the stretch. Inside of our function we have $+2$; this tells us we start by shifting left 2 units (solid):

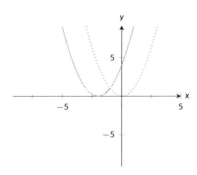

Next, we will stretch our graph horizontally (solid) by a factor of 2 since x is multiplied by $\frac{1}{2}$ on the inside of the function:

Notes:

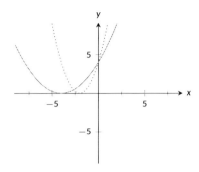

Now that we have completed all horizontal changes, we can work on vertical changes. We have two: a flip and a shift. We need to do the flip first (solid):

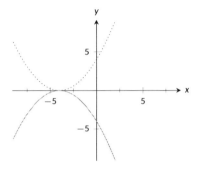

The last modification to complete is a shift up 3 units (solid):

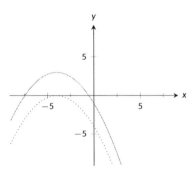

In all of our examples so far, we've been given the transformed function and asked to graph it. What if we are asked to come up with the new function?

Notes:

Example 59   **Functions for a Transformation**

Determine the equation for the graph of $f(x) = x^3$ after it has been shifted 2 units to the right, flipped vertically, and shifted 2 units up.

Solution   Here the transformations have been given in the same order that we would apply them. Our first step, is to shift the function to the right, so we will change to $(x-2)^3$. Next, we want to flip the function vertically, so we get $-(x-2)^3$. Finally, we want to shift up 2 units, so we get $g(x) = -(x-2)^3 + 2$ as our new function. we gave the function a new name g, instead of f so that we won't get confused by using the same name for both.

Graphs of Piecewise Functions

The last graphing topic we will discuss in this section is graphing piecewise functions. A piecewise function is a function that is defined in pieces; for part of its domain it is defined one way and for other parts it is defined differently. When graphing these functions, the trickiest part is making sure you use the correct piece of the function definition for each part of the domain. To make this a little easier to keep straight, we make sure to only graph one piece at a time.

When switching between different pieces, it is important to do so properly. Sometimes the "end" point of that piece is not actually included. This happens when the domain for that piece is open, i.e., it doesn't include that final point. This is indicated by a parenthesis or a "$<$" or a "$>$" telling us that the end point should not be included. Here, we would plot an open dot, a circle with a white interior, to show that it is not included. If the endpoint is included, we will plot a closed dot, a circle with a filled in interior, to show that it is included. If the function's domain for a piece continues all the way to ∞ or to $-\infty$, we will indicate this by drawing a small arrow tip at the edge of the graph to show that it continues forever.

Example 60   **Graphing a Piecewise Function**

Graph the function $f(x) = \begin{cases} 3-x & 0 < x \leq 1 \\ x^2 & 1 < x < 2 \end{cases}$

Solution   This function has two pieces: a line when x is in $(0, 1]$ and a quadratic function when x is in $(1, 2)$. We like to work from left to right, so we will graph the line first, but you could graph the pieces in any order.

Notes:

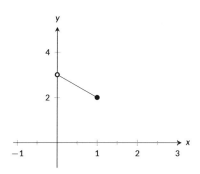

Here, we graphed the line by plotting the two end points and connecting them. Since they are the endpoints, we don't want to move past them. Since the line applied when $0 < x \leq 1$, we plotted an open dot for $x = 0$ to show that it is not included and a closed dot for $x = 1$ to show that it is included.

Next, we'll add the quadratic piece to this graph.

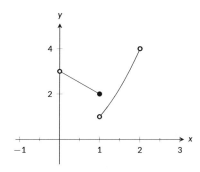

With the quadratic, both ends are open since this piece only applies when $1 < x < 2$. This means we need open dots at both ends.

Notes:

Exercises 2.4

Terms and Concepts

1. Why do changes on the inside of the function produce horizontal changes?

2. Why do changes on the outside of the function produce vertical changes?

3. In graphing the function $g(x) = 2\ln(x) + 4$, which transformation should you apply first?

4. In graphing the function $f(x) = (2x - 1)^3$, which transformation should you apply first?

5. In graphing the function $h(t) = 3^{t+4}$, what is the base function and how is it being transformed?

Problems

Graph each of the functions in exercises 6 – 10.

6. $f(x) = -x^2 + 1$

7. $f(x) = \begin{cases} 2x + 8 & x \leq -1 \\ -x + 7 & x > -1 \end{cases}$

8. $f(x) = \begin{cases} -x^2 & x < 0 \\ (x - 1)^2 & 0 \leq x < 3 \end{cases}$

9. $f(x) = e^x + 1$

10. $f(x) = \begin{cases} \sin(x) & x < \pi \\ \cos(x) & x > \pi \end{cases}$

In exercises 11 – 15, graph and write an equation for each of the described functions.

11. The result of shifting $g(x) = x^2$ up three units and to the left two units

12. Any even degree polynomial that is positive for $-2 \leq x \leq 4$.

13. The result of shifting $f(\theta) = -2\theta + 3$ down two units and right 5 units.

14. The piecewise function that consists of t^2, shifted down one unit for $t \leq -2$ and of the line with a slope of 3 and a y-intercept of 3 for $t > -2$.

15. The line with a slope of $\frac{2}{3}$ that passes through the point $(1, f(2))$, where $f(x) = x^2 - 1$.

In exercises 16 – 18, factor the given function, and graph the function.

16. $b(x) = x^3 + 6x^2 + 12x + 8$

17. $y(t) = t^2 - 6t + 9$

18. $f(x) = x^2 + 4x + 4$

For each of

- $f(x) = x^2 - 3x$,

- $\eta(\theta) = \cos(\theta)$, **and**

- $g(w) = 3^w - w^3$,

write the equation for the new function that results from the transformation(s) stated in exercises 19 – 24.

19. Shift up 3 units

20. Shift right 2 units

21. Shift down 2 units and left 1 unit

22. Shift down π units and right e units

23. Flip across the x-axis

24. Flip across the y-axis

Answer each question in exercises 25 – 26 using the provided graphs.

25. Based on the shape of the graph of $f(x)$ shown, below,

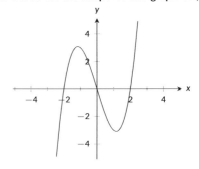

(a) could $f(x)$ be an even polynomial? Why or why not?

(b) could $f(x)$ be an odd polynomial? Why or why not?

(c) could $f(x)$ be an exponential function? Why or why not?

(d) could $f(x)$ be a trigonometric function? Why or why not?

26. Based on the shape of the graph of $g(x)$ shown, below,

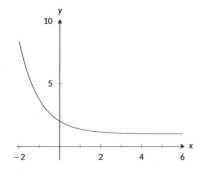

(a) could $g(x)$ be an even polynomial? Why or why not?

(b) could $g(x)$ be an odd polynomial? Why or why not?

(c) could $g(x)$ be an exponential function? Why or why not?

(d) could $g(x)$ be a trigonometric function? Why or why not?

2.5   Completing the Square

In this section, we will discuss another way of writing a quadratic function through a process called completing the square. Completing the square lets us write any quadratic function in the form $(x + a)^2 + b$. This particular form is quite handy; not only will this making graphing quadratics easier since it allows us to use graph transformations, it's also a commonly used form in calculus. In integral calculus, there are special rules that allow us to more easily integrate rational functions if their denominator is in the form, but we aren't always given the function in that form. It also gets used when working with Laplace Transforms in differential equations. Since it appears so frequently in later courses, it's a good idea to master this skill now.

The ideas behind the techniques we will use for completing the square build off of our ideas from expanding. We looked at common patterns, and one of the ones we discussed was

$$(u + v)^2 = u^2 + 2uv + v^2$$

We will use this pattern to help us change quadratic functions of x into the form $(x + a)^2 + b$. If we use this expansion pattern, we see that

$$(x + a)^2 = x^2 + 2ax + a^2 \tag{2.6}$$

We will use the coefficient on the x term of our quadratic function to help us find a. Once we have a, we can calculate a^2 and use it to help us determine what b needs to be to write our quadratic in the form $(x + a)^2 + b$. Let's give it a try.

Example 61   **Completing the Square**
Write $f(x) = x^2 + 4x + 6$ in the form $(x + a)^2 + b$.

Solution   We saw in equation 2.6 that $(x + a)^2 = x^2 + 2ax + a^2$. We'll use the coefficient on x from our function to determine a.

In $f(x)$, x has a coefficient of 4 and in the expanded pattern x has a coefficient of $2a$. We want these to match, so we get $4 = 2a$, or $a = 2$. Let's see what our pattern looks like with $a = 2$:

$$(x + 2)^2 = x^2 + 4x + 4$$

This is pretty close to what $f(x)$ looks like; the only difference is the constant term. Remember, our end goal is to write $f(x)$ in the form $(x + a)^2 + b$. We've already figured out a; now we need to figure out b. With the addition of b, we can expand our goal form to get

$$(x + a)^2 + b = x^2 + 2ax + a^2 + b$$

Notes:

This tells us that b influences our constant term. We want the constant terms to match, so we have $6 = a^2 + b$. We know $a = 2$, so really we have $6 = 4 + b$, giving us that $b = 2$. That means we have

$$f(x) = (x + 2)^2 + 2$$

This example shows the line of thinking we used with this problem, but is a fairly wordy explanation. Mathematicians like to keep things concise, so let's see how we could show this work mathematically, without using much of a verbal description. Typically, you will see work like this:

$$\begin{aligned}
f(x) &= x^2 + 4x + 6 \\
&= x^2 + 2(2x) + 6 \\
&= x^2 + 2(2x) + (2^2) - (2^2) + 6 \\
&= (x + 2)^2 - (2^2) + 6 \\
&= (x + 2)^2 - 4 + 6 \\
&= (x + 2)^2 + 2
\end{aligned}$$

(2.7)

This work shows the same steps we did above, but in a different form, and without explicitly saying what a and b are. However, you can see that these steps are working towards the form we want by using our pattern. In the second line, we write $2(2x)$ instead of $4x$ to figure out a. Then, since we know we have a^2 as part of our constant, we add 2^2 and subtract 2^2 in the same step. Why? Well, this makes sure we add zero, that we don't change the meaning of the function, just the way it's written. Then, we have the correct pattern to write the first three terms as $(x + 2)^2$. Lastly, we simplify the constants outside of the parentheses to find b.

In practice, most mathematicians may combine a couple of the steps into one, but until you really get comfortable with this line of thinking it's best to write out all the steps.

Most people learn completing the square as an algorithm, a set of steps that must be performed exactly as described and in the correct order to get the final answer. We are intentionally avoiding such an algorithm here; algorithms can be difficult to memorize, but are easy to forget. If you instead think of this as a puzzle where you figure out one part at a time, it's more likely that you will still be able to accurately complete the square in later courses.

We've looked at one example of completing the square that had all "nice" numbers in it, now let's take a look at one that's a bit messier.

Notes:

Example 62   **Completing the Square**
Complete the square for $g(t) = t^2 - 7t + 10$.

Solution   There is one big difference between this problem and our previous example: our input variable has changed. That means instead of our goal looking like $(x + a)^2 + b$, our goal looks like $(t + a)^2 + b$. Regardless, we'll follow the same thought process we used in the previous example. We know that if we expand our goal form, we get $(t + a)^2 + b = t^2 + 2at + a^2 + b$. Like before, we'll figure out a value for a first, and then a value for b. To find a, we will use the t term. The expanded goal form has $2at$ and $g(t)$ has $-7t$. This tells us that $2a = -7$, or $a = -\frac{7}{2}$. In the expanded goal form, the constant term is $a^2 + b$; we know a now, so really we have $\frac{49}{4} + b$. Note that when we square a we get a positive number (think back to the invisible parentheses we talked about earlier). In $g(t)$, our constant term is 10. Matching our constant terms gives us the equality $\frac{49}{4} + b = 10$. If we subtract $\frac{49}{4}$ from both sides, we get $b = -\frac{9}{4}$.

Altogether, we have $a = -\frac{7}{2}$ and $b = -\frac{9}{4}$, so we have

$$g(t) = \left(t - \frac{7}{2}\right)^2 - \frac{9}{4}$$

A Variation on Completing the Square

In all of the examples we have discussed in this section, the squared term has a coefficient of 1. However, sometimes we will run into situations where this coefficient isn't 1, and we will need to be able to work with these situations. When a quadratic in x (meaning a quadratic function that has x as its variable) has a leading coefficient (the coefficient on the highest power term) other than 1, we can write it as $c(x+a)^2 + b$. This means that we will have three parameters we need to find: a, b, and c. In our earlier examples we found a first because only it showed up on the x term, and the x^2 term was already taken care of since it automatically had a coefficient of 1. Here, we will want to find c first since it shows up in the quadratic term and affects the linear term and the constant term. This is a common solution technique in mathematics: start by working with the highest power terms first, and then move onto the lower degree terms. Before we look at an example problem, let's see what this modified form looks like after expansion. We have

---

Notes:

$$c(x + a)^2 + b = c(x^2 + 2ax + a^2) + b$$
$$= cx^2 + 2acx + ca^2 + b$$

$$(2.8)$$

There are some key features we need to note that will be handy when dealing with these types of quadratics. In this form c impacts the x^2, x, and constant terms. For this form, we will start by "matching" coefficients with the x^2 term, then the x term, and then the constant term. Let's take a look:

Example 63   **Completing the Square- Variation**
Write $f(x) = 4x^2 + 12x - 3$ in the form $c(x + a)^2 + b$.

Solution   Since we want our answer in the form $c(x + a)^2 + b$, we will use equation 2.8. In equation 2.8, we see that the coefficient on x^2 in the expanded form is c. For $f(x)$, the x^2 coefficient is 4, so we have $c = 4$.

Next, we'll work with the x term. In equation 2.8, the x term has a coefficient of $2ac$. We are using $c = 4$, so really this coefficient is $2a(4) = 8a$. For $f(x)$, the x coefficient is 12, so we get $8a = 12$, or $a = \frac{3}{2}$.

Lastly, we'll work with the constant terms. In equation 2.8, the constant is $ca^2 + b$. Since we have $a = \frac{3}{2}$ and $c = 4$, this constant really is $4(\frac{3}{2})^2 + b = 9 + b$. In $f(x)$, the constant is -3, so we have $9 + b = -3$, or $b = -12$.

We've now found all three parameters, so we are done and have that

$$f(x) = 4\left(x + \frac{3}{2}\right)^2 - 12$$

We could also solve this problem a bit differently. We could start by factoring out the coefficient on the x^2 term and then completing the square on what remains. Let's take a look:

Example 64   **Completing the Square- Variation**
Write $f(x) = 4x^2 + 12x - 3$ in the form $c(x + a)^2 + b$.

Solution   We'll start by factoring 4 out from the equation and completing the square on the remaining quadratic factor. By factoring out the 4, x^2

Notes:

will have a coefficient of 1 and we can work like we did in our earlier examples.

$$f(x) = 4x^2 + 12x - 3 = 4\left[x^2 + 3x - \frac{3}{4}\right]$$

$$= 4\left[x^2 + 2\left(\frac{3}{2}\right)x + \left(\frac{3}{2}\right)^2 - \left(\frac{3}{2}\right)^2 - \frac{3}{4}\right]$$

$$= 4\left[\left(x + \frac{3}{2}\right)^2 - \frac{9}{4} - \frac{3}{4}\right]$$

$$= 4\left[\left(x + \frac{3}{2}\right)^2 - \frac{12}{4}\right]$$

$$= 4\left[\left(x + \frac{3}{2}\right)^2 - 3\right]$$

We're close to the form we want, but we have an extra set of parentheses. We will need to redistribute the 4 to the rest of the statement to be in the correct form. This gives us

$$f(x) = 4\left(x + \frac{3}{2}\right)^2 - 12$$

As you can see, we end up with the exact same answer either way, but used a different method. With the first method, we expanded the general form we wanted and found the values of a, b, and c one by one. With the second method, we started with our specific function $f(x)$, and rearranged it to look like the form we want. With the second method, the values of a, b, and c can be identified from our final answer.

When trying to rewrite a function into a different form, it's very important to pay close attention to how the form is written, particularly if that form is used as part of a rule that you need to fully solve the problem you are working on. The parameters may not always be in alphabetical order and mixing up the parameter values could drastically change your final answer. Additionally, some books do not always use the same letters in the same positions, even if it's the same rule. Many rules will reuse the same letters as parameters, but they quite often are filling different roles.

Notes:

Exercises 2.5

Terms and Concepts

1. How would completing the square on a quadratic function help you graph it?

2. After completing the square, you get $f(x) = (x - 2)^2 + 3$. Is $x = 2$ considered a root of $f(x)$? Explain.

3. One of the variations on completing the square gives you the form $(cx + a)^2 + b$. Does c represent a vertical stretch/shrink or a horizontal stretch/shrink of the function x^2?

4. After completing the square you get that $g(t) = (t+2)^2 - 6$. What are the values of a and b if your goal form is $(x+a)^2 + b$?

Problems

In exercises 5 – 11, write each function in the form $(x+a)^2 + b$ and identify the values of a and b.

5. $f(x) = x^2 - 4x + 6$

6. $g(x) = x^2 + 20x + 40$

7. $h(x) = x^2 - 8x + 5$

8. $m(x) = x^2 - 22x - 4$

9. $n(x) = x^2 - 6x - 2$

10. $p(x) = x^2 + 11x + 4$

11. $p(x) = x^2 + 13x$

In exercises 12 – 16, write each function in the form $c(x + a)^2 + b$ and identify the values of a, b, and c.

12. $f(x) = 9x^2 - 12x + 12$

13. $g(x) = x^2 - 2x + 2$

14. $h(x) = 4x^2 - 4x - 4$

15. $w(x) = 4x^2 + 4x + 6$

16. $y(x) = 9x^2 + 18x + 4$

In exercises 17 – 20, complete the square and use your result to help you graph the function.

17. $f(t) = t^2 + 2t + 3$

18. $p(q) = q^2 - \frac{2}{3}q$

19. $y(x) = x^2 + 4x + 2$

20. $f(x) = x^2 - 4x + 6$

In exercises 21 – 24, expand and graph the function.

21. $f(x) = (x - 1)^2 - 2$

22. $g(x) = -(x + 3)^2 + 4$

23. $h(x) = (-x + 3)^2 + 4$

24. $x(y) = (y + 2)^2 - 1$

3: Solving and Trigonometric Functions

In this chapter, we will look at some special types of functions that are commonly used in calculus: trigonometric functions. Additionally, we will look at solving complicated equations for a given variable and finding all of the points where two functions intersect each other. Each of these skills are quite important in calculus. Trigonometric functions help model natural phenomena such as sound and light waves, and are used in related rates to determine how quickly something, like an angle, is changing. Because of the varied applications you will see in calculus, familiarity with these functions is a must. We will also look at a way that we can take a rational function and write it in a different form. Sometimes, one of these forms will be more useful to us than another form, particularly in integral calculus where we will have rules that only work for certain function forms. When we look at the intersections of two functions, we will mostly focus on polynomials in this chapter since they are commonly used functions in scientific fields. Intersections will be used frequently in integral calculus when we are determining the area enclosed by two or more functions.

3.1   Solving for Variables

In many math and science courses, you will need to be able to isolate, or solve for, a variable. This skill gets used in many places: in differential calculus it will help you identify the maximum or minimum value of a function and when you perform implicit differentiation where you will need to express one variable in terms of several variables and parameters; in integral calculus you will use it when you work to identify intersection points, and in the many physics based problems you will see in differential calculus such as equations working with spring motion. In many of these situations, you will either have multiple variables, lots of parameters, or fairly complicated functions. In all of these situations, the first step to solving for a variable involves figuring out what type of expression or function that you are working with. The type of expression guides the solution process; we take a different approach for quadratics than we do for linear expression and a different approach yet for trigonometric expressions. We will learn some of these approaches in this section, but we won't discuss approaches for trigonometric functions until we discuss these functions in greater detail, and we've already seen how to solve logarithmic and exponential functions (see Section 1.5).

When identifying the type of expression, you'll need to make sure you are focusing on your variable of interest. For example, we've looked at a formula for the height of a ball that has been thrown into the air, formula 1.14:

$$h(t) = h_0 + v_0 t + \frac{1}{2} a t^2$$

We discussed how this function has one variable, t, and several parameters: h_0, v_0, and a. Here, since t is our variable, we naturally focus on t and say that we have a quadratic function of t. However, there may be situations where our focus shifts. For example, you may want the ball to have a certain height, say 10 meters, after $t = 30$ seconds and you are able to adjust the initial velocity, v_0, that the ball is thrown with. In this situation our focus is really on v_0 and not on t. This would give us the equation

$$10 = h(30) = h_0 + v_0(30) + \frac{1}{2} a(30)^2, \text{ or}$$

$$10 = h_0 + 30v_0 + 450a$$

(3.1)

Here, our equation still includes the parameters h_0, v_0, and a, but we've substituted 30 for t, and we are looking for the value of v_0 that makes this statement true. Since we are focusing on v_0, we really only have a linear equation of v_0. In general, we would say that $h(t)$ is linear in v_0, meaning that if everything else is treated like a parameter, the highest degree of v_0 is 1, so it is linear. Similarly, $h(t)$ is a linear function of h_0 and a linear function of a. Let's take a look at a few more examples.

Example 65   Determining Statement Type
The statement $y^3 \sin(w) = 4y^2 z + 2x^2 y + xy + 10$ is what type of statement in terms of

1. the letter w?

3. the letter y?

2. the letter x?

4. the letter z?

Solution

1. Here, our focus is on the letter w. The letter w only appears on the left-hand side in the term $y^3 \sin(w)$. Since w is inside of the sine function,

| This statement is trigonometric in w |

2. Here, our focus is on the letter x. The letter x appears in two terms: $2x^2 y$ and xy. The remaining terms, $y^3 \sin(w)$, $4y^2 z$, and 10 do not involve x, so they are considered to be constant terms when we focus on x. Out of the terms that include x, we have an x^2 term and an x term. This tells us that

Notes:

The statement is quadratic in x

3. Now, we are focusing on y. The letter y shows up in every term except 10. We have a y^3 term, a y^2 term, two y terms, and a constant term. We could rewrite the statement so that there only looks like one y term by gathering like terms and rewriting $2x^2y + xy$ as $(2x^2 + x)y$. Since we only have these four different types of terms,

This statement is cubic in y

4. Lastly, we'll look at z. The letter z only shows up in the term $4y^2z$; every other term counts as a constant when we focus on z. This tells us that

The statement is linear in z

Note that in the previous example, we do not know if each letter is representing a variable or a parameter because we are given no context for the statement. Since we are unsure, we just referred to "letters" so that we did not add meaning that may not be correct.

Identifying the type of statement or equation for our variable of interest is our first step in solving for the variable. Next, we'll look at how to solve when we have linear statements, quadratic statements, or special kinds of higher degree polynomials. We'll also discuss the first steps for solving trigonometric functions, but we won't learn how to fully solve these until later sections.

Solving a Linear Statement

Solving a linear statement is rather straightforward. For this explanation, we'll use x as our variable of interest. Start by moving every term that doesn't have an x to one side, and every term that does have an x to the other side of the statement. Then, factor the x out of every term with x. Lastly, divide the side without x by the coefficients on x. Let's take a look at solving a linear statement.

Example 66   **Solving a Linear Statement**
Solve the statement $xz + 2yz - 4 = \sin(x) + y^2 + y^2z$ for z.

   Solution   Here we see that we do have a linear statement in z: the highest degree of z in the statement is 1, and z does not appear inside of any

---

Notes:

other functions. We'll start by gathering every term with a z on the left side by subtracting y^2z from both sides:

$$xz + 2yz + y^2z - 4 = \sin(x) + y^s$$

Next, we'll gather all terms that don't have a z on the right side:

$$xz + 2yz + y^2z = \sin(x) + y^2 + 4$$

We did this by adding 4 to both sides. Next, we'll factor out z on the left side since it is a common factor for all of those terms:

$$z(x + 2y + y^2) = \sin(x) + y^2 + 4$$

Lastly, we'll divide the right side by the z coefficient on the left side:

$$z = \frac{\sin(x) + y^2 + 4}{x + 2y + y^2}$$

Solving a Quadratic Statement

When we solve quadratic statements, we'll build off of the skills we learn with factoring quadratics and with finding the roots of quadratics. We know that if we want to find the roots of a function, we are really looking for all inputs that give us an output of zero. So, we take the function and set it equal to zero and use the quadratic formula. We know we could instead find the factors of the function and use those to find the roots. Here, we will focus on using the quadratic formula since we are dealing with fairly complicated expressions that will be more difficult to factor. With either method, we are building off of this idea of finding roots/factors, which relies on a statement where one side is zero. This means that when we work with quadratic statements, we will need to move all of our terms to one side before we do anything else. This is a common tripping point for people working with quadratics. Rather than developing a whole new set of techniques, mathematicians like to use old techniques as much as possible, and here that means we need one side to be zero.

Once we've moved all terms to one side, then we can gather our like terms. We know that our quadratic formula relies on knowing the coefficients of the x^2 term, the x term, and the constant term, so we will want to gather all x^2 terms together, to gather all x terms together, and to gather all constant terms together. From each set, we will factor out the x's to find these coefficients. Let's take a look.

Notes:

Example 67   **Solving a Quadratic Statement**
Solve the statement $xz + 2yz - 4 = \sin(x) + y^2 + y^2z$ for y.

Solution   We saw this statement in our last example, but there we were solving for z. Here, we want to solve for y. We can see that this statement is quadratic in y because we only have y^2 terms, y terms, and constant terms. We'll start by moving every term on the right to the left (you could move everything to the right instead; you would end up with the same final answer).

$$xz + 2yz - 4 - \sin(x) - y^2 - y^2z = 0$$

Now, we will rearrange the order of our terms to gather all of the like terms. We'll start with the y^2 terms, then the y terms, then the constant terms.

$$-y^2 - y^2z + 2yz + xz - 4 - \sin(x) = 0$$

Now we will factor out the y^2 from the first two terms, then the y from the next term.
$$y^2(-1 - z) + y(2z) + xz - 4 - \sin(x) = 0$$

Notice that when we factored y^2 out of the first two terms, we did not factor out the negative, even though both terms are negative. This is because we want the negative to be part of the y^2 coefficient in order to be in the right form to use the quadratic formula. Now, we can use the quadratic formula. In the quadratic formula, a is the coefficient on the squared term, so here we have $a = -1 - z$. Next, b is the coefficient on the y term, so $b = 2z$. Lastly, c is the constant term, so we have $c = xz - 4 - \sin(x)$. Substituting into the quadratic formula, we get:

$$y = \frac{-b \pm \sqrt{b^2 - 4ac}}{2a}$$

$$= \frac{-(2z) \pm \sqrt{(2z)^2 - 4(-1 - z)(xz - 4 - \sin(x))}}{2(-1 - z)}$$

$$= \frac{-2z \pm \sqrt{4z^2 - 4(-1 - z)(xz - 4 - \sin(x))}}{-2 - 2z}$$

$$= \frac{-2z \pm \sqrt{4z^2 + (4 + 4z)(xz - 4 - \sin(x))}}{-2 - 2z}$$

$$= \frac{-2z \pm \sqrt{4z^2 + 4xz - 16 - 4\sin(x) - 4xz^2 - 16z - 4z\sin(x)}}{-2 - 2z}$$

Since this cannot be easily simplified, we will leave the answer as it is, giving us a final answer of

Notes:

$$y = \frac{-2z \pm \sqrt{4z^2 + 4xz - 16 - 4\sin(x) - 4xz^2 - 16z - 4z\sin(x)}}{-2 - 2z}$$

As you probably noticed, the answer to this example is quite complicated, which shows us that trying to find factors for this quadratic, rather than the roots, is quite difficult. It's important to notice that we get two answers here:

$$y = \frac{-2z + \sqrt{4z^2 + 4xz - 16 - 4\sin(x) - 4xz^2 - 16z - 4z\sin(x)}}{-2 - 2z}$$

$$y = \frac{-2z - \sqrt{4z^2 + 4xz - 16 - 4\sin(x) - 4xz^2 - 16z - 4z\sin(x)}}{-2 - 2z}$$

We get two answers because of the $\pm$; this tells us that one answer comes from using $+$ and the other from using $-$. This shows us that every quadratic statement will have two solutions. Sometimes these solutions will look the same (if we solve $(x-1)^2 = 0$ both answers will be $x = 1$), and then mathematicians say there is *one repeated solution* (or one repeated root). This repetition of the root does make a difference in other properties of the function that you will discuss in differential calculus. For the previous example, if the square root evaluates to zero, we would have one repeated solution. If we end up with the square root of a positive number, we would have *two distinct (different) solutions* (or roots), and if we end up with the square root of a negative number, we would have a *complex conjugate pair of solutions* (roots). Complex means that our answers involve imaginary numbers (any number that involves $i = \sqrt{-1}$) and conjugate pair means that they are related, and the only difference is that for one answer we use the $+$ part of the $\pm$ and for the other we use the $-$ part of the $\pm$. This shows that for this example, our final answer could change drastically depending on the values of x and z.

For quadratics, these are our only options for classifying our answers. In later courses like differential equations, the type of answer will tell you about properties of related functions. Notice that for a quadratic, we have 2 answers and the degree of the statement is 2. For linear statements, we only have one answer, and the degree of the statement is 1. This pattern continues; for cubic statements you have 3 answers and the degree is 3, for quartic statements you have 4 answers and the degree is 4. With any polynomial with degree of 2 or more, we could have solutions that are repeated, or complex conjugate pairs, as well as distinct solutions. The complex conjugate pairs give you two answers; for a cubic if you have a complex conjugate pair of solutions, your other solution must be a real solution. Let's take one look at a quadratic that has complex

Notes:

conjugate solutions so you can see how to write your answer using i, which is the most common way to express these answers.

Example 68   **Solving a Quadratic Statement**
Solve $x(x - 4) = -13$ for x.

Solution   From the form of the statement we are given, it's not entirely obvious that we are working with a quadratic. We'll need to expand the left side before we do anything else so that we can easily see that it is a quadratic and so we can correctly determine all of the coefficients. Expanding gives $x^2 - 4x = -13$. Our next step is to move everything to one side:

$$x^2 - 4x + 13 = 0$$

Now, we can pick out our values of a, b, and c for the quadratic formula. x^2 has a coefficient of 1, so $a = 1$; x has a coefficient of -4, so $b = -4$; and the constant is -13, so $c = -13$. The quadratic formula gives us

$$x = \frac{-b \pm \sqrt{b^2 - 4ac}}{2a}$$

$$= \frac{-(-4) \pm \sqrt{(-4)^2 - 4(1)(13)}}{2(1)}$$

$$= \frac{4 \pm \sqrt{16 - 52}}{2}$$

$$= \frac{4 \pm \sqrt{-36}}{2}$$

Now, we'll need to deal with that negative under the square root before we simplify our answer. We can deal with this by factoring the square root: $\sqrt{-36} = \sqrt{(-1)(36)} = \sqrt{-1}\sqrt{36} = i\sqrt{36}$. Now, we can continue simplifying our answers:

$$x = \frac{4 \pm \sqrt{-36}}{2}$$

$$= \frac{4 \pm i\sqrt{36}}{2}$$

$$= \frac{4 \pm i(6)}{2}$$

$$= \frac{4 \pm 6i}{2}$$

$$= 2 \pm 3i$$

Notes:

So, we end up with a complex conjugate pair:

$$x = 2 + 3i \text{ and } x = 2 - 3i$$

Solving Higher Degree Statements

For higher degree polynomial statements that do not include parameters, we can build off of the strategies we used when factoring and finding roots. For these, we can solve by moving all of the terms to one side, and then finding the roots. These roots will be the same as the solutions to the original statement. However, solving higher degree statements can be quite difficult if they have many parameters. For cubic functions, there is a formula (akin to the quadratic formula) that will allow you to solve if you move all terms to one side, but the formula is long and hard to simplify. Few mathematicians could tell you this formula without looking it up because it is rarely used; because of this we will not cover it, but it is good to know that it exists. Similarly, an even more complicated formula exists to find a root of a quartic function; once you find the first root you would then have to use the cubic function formula. For any functions of degree 5 or higher, there is no known formula to help you solve. Since these higher degree functions are quite difficult to work with, we will not work with them. We have, however, already looked at solving special cases of them when we discussed exponents in Section 1.4.

Solving Non-polynomial Statements

Lastly, we will briefly discuss solving non-polynomial statements, statements where our letter of interest is an input for a trigonometric function, a logarithmic function, or an exponential function. Trigonometric statements require the use of tools and ideas that we have not yet discussed, so the full details will be covered in later sections. However, for all of these, the first half of the process is the same as solving for a linear statement. We start by isolating the function that involves our letter of interest by moving all other terms to the other side, and then dividing by any coefficients on our term of interest. For example, if we wanted to solve the statement $\ln(a)b^3cx - 7 + \sin(b) = 2x^2$ for a, we would start by isolating $\ln(a)$, and would end up with $\ln(a) = \dfrac{2x^2 + 7 - \sin(b)}{b^3cx}$. This doesn't tell us what a is, but we're nearly there. We've already seen how to solve for a, now that $\ln(a)$ is isolated, in Section 1.5. Now, you have the tools you need to solve most statements you will encounter in calculus.

Notes:

Exercises 3.1

Terms and Concepts

1. Is it possible to solve a cubic statement?

2. What are the possible types of solutions when solving a quadratic statement?

3. What is the maximum number of different solutions that a seventh degree statement could have?

4. T/F: A cubic statement can have only complex solutions. Explain.

Problems

In exercises 5 – 10, determine the type of statement in terms of the given variable.

5. $x^3y + 2x^2yz - 6xz^2 = yz^2 - 10$ in terms of x

6. $x^3y + 2x^2yz - 6xz^2 = yz^2 - 10$ in terms of y

7. $x^3y + 2x^2yz - 6xz^2 = yz^2 - 10$ in terms of z

8. $xt + \cos(\theta) = x^4t^3 - 6t$ in terms of θ

9. $xt + \cos(\theta) = x^4t^3 - 6t$ in terms of x

10. $xt + \cos(\theta) = x^4t^3 - 6t$ in terms of t

In exercises 11 – 18, determine if it is possible to solve the statement for the given variable. If it is possible, solve but do not simplify your answer(s). If it is not possible, explain why.

11. $xy^2 - xy = 5y - 3x$ for x

12. $xy^2 - xy = 5y - 3x$ for y

13. $3t^2 - 5mq = 8qt + 2m^3$ for q

14. $2a^2bc^3 + 3abc^2 + 4a^2c^2 - 3b = 4c$ for a

15. $2a^2bc^3 + 3abc^2 + 4a^2c^2 - 3b = 4c$ for b

16. $\log_2(xy) = x + e^z$ for x

17. $\log_2(xy) = x + e^z$ for y

18. $\log_2(xy) = x + e^z$ for z

In exercises 19 – 27, solve for x. Be sure to list all possible values of x.

19. $x^2 - 16 = 0$

20. $x^2 + 16 = 0$

21. $x^2 - 4x - 7 = 2$

22. $x^2 - 2x + 7 = 2$

23. $5x^2 + 2x = -1$

24. $x^3 = 8$

25. $x^3 + x^2 = 4x + 4$

26. $2(x - 3)^2 - 7 = -4x + 9$

27. $(x + 2)^3 = 2x^2 + 8x + 7$

In exercises 28 – 32, Classify the type(s) of solution(s) from the given exercise.

28. Exercise 19

29. Exercise 20

30. Exercise 21

31. Exercise 24

32. Exercise 25

3.2   Intersections

In many problems in integral calculus you will be finding the area enclosed by, or between, several functions. As part of finding the area, you will need to identify where the functions intersect each other, i.e., the (x, y) coordinate pairs where the curves cross. The *points of intersection* of two functions, $f(x)$ and $g(x)$, are the (x, y) coordinate pairs for which the input, x, results in the same output value from both functions. In this section, we will address three different methods for finding the points of intersection for two graphs. The first two methods we will discuss rely heavily on this skills you learned in the previous section where you learned how to solve for variables.

Note that while we have mostly been using function notation like $f(x)$, here we will often indicate the output of the function as y. One reason why we are using y here is that some of our functions will be defined *implicitly*. When a function is defined implicitly, it means that the output of the function is not isolated; we've seen this before with the point slope form of a line. When the output is isolated, we say our function is defined *explicitly*, as in slope intercept form.

Substitution

Substitution is most commonly used when one or both functions are defined implicitly, or when both functions have a term in common. With this method, we will solve one equation for one of the variables and then substitute the solution into the second equation and solve for the remaining variable. In this course, we are only interested in real number solutions. Here, it is a matter of personal preference when choosing which function to work with initially, and which variable to solve for. However, we recommend starting with the equation that is "simpler;" if one equation is linear and the other is quadratic, it is typically less complicated to start with the linear function. Let's take a look at an example.

Example 69   **Points of Intersection: Substitution**
Find the points of intersection for $4x^2 + y^2 = 4$ and $y - 1 = 2(x - 1)$.

Solution   Here, the first equation, $4x^2 + y^2 = 4$ is quadratic in both x and in y, but the second equation, $y - 1 = 2(x - 1)$ is linear in both x and in y. Because of this, we will start our work with the second equation.

Additionally, in the second equation y is already nearly isolated, so we will first isolate y in this equation.

Notes:

$$y - 1 = 2(x - 1)$$
$$y = 2(x - 1) + 1$$
$$= 2x - 2 + 1$$
$$= 2x - 1$$

Now that we have y isolated, we will replace every y in the equation $4x^2 + y^2 = 4$ with $2x - 1$. As when we evaluated functions, we will be sure to put parentheses around the term, $(2x - 1)$, so that we can simplify correctly.

$$4x^2 + y^2 = 4$$
$$4x^2 + (2x - 1)^2 = 4$$
$$4x^2 + (4x^2 - 4x + 1) = 4$$
$$8x^2 - 4x - 3 = 0$$

This doesn't look like it is likely to factor nicely, so we will use the quadratic formula:

$$x = \frac{-(-4) \pm \sqrt{(-4)^2 - 4(8)(-3)}}{2(8)}$$

$$= \frac{4 \pm \sqrt{16 + 96}}{16}$$

$$= \frac{4 \pm \sqrt{112}}{16}$$

$$= \frac{4 \pm 4\sqrt{7}}{16}$$

$$= \frac{1 + \sqrt{7}}{4}, \frac{1 - \sqrt{7}}{4}$$

This only gives us the x coordinates; we also need the y coordinates. To get the corresponding y coordinates, we will use the linear equation where we already solved for y in terms of x. We could use the earlier form of this equation, or we could even use the quadratic equation, but either of these would require more work. The first y coordinate is:

$$y = 2x - 1$$

$$= 2\left(\frac{1 + \sqrt{7}}{4}\right) - 1$$

Notes:

$$= \frac{1 + \sqrt{7}}{2} - \frac{2}{2}$$

$$= \frac{-1 + \sqrt{7}}{2}$$

The second y coordinate is:

$$y = 2x - 1$$

$$= 2\left(\frac{1 - \sqrt{7}}{4}\right) - 1$$

$$= \frac{1 - \sqrt{7}}{2} - \frac{2}{2}$$

$$= \frac{-1 - \sqrt{7}}{2}$$

Now, we have both of the points of intersection:

$$\left(\frac{1 + \sqrt{7}}{4}, \frac{-1 + \sqrt{7}}{2}\right) \text{ and } \left(\frac{1 - \sqrt{7}}{4}, \frac{-1 - \sqrt{7}}{2}\right)$$

Sometimes, we can be a bit creative about using substitution. Depending on the equations you are working with, it may sometimes be quicker to *not* solve for a variable completely, but rather for a term that shows up in both equations. Let's take a look at an example.

Example 70   **Points of Intersection: Substitution**
Find all points of intersection of $x - 4 = y^2$ and $x^2 - 4x = -y^2$.

Solution   Here we can see that the only "easy" place to start by solving would be to solve for x in the first equation, but once we substitute into the second equation, things will get messy quickly. However, both equations have a y^2 term, and no other y terms. This means that we can save some work by solving for y^2 in one equation and substituting into y^2 in the other equation. Since the first equation already has y^2 isolated, we really just have to do the substitution. We will substitute $x - 4$ into the second equation in the place of y^2:

$$x^2 - 4x = -y^2$$

$$x^2 - 4x = -(x - 4)$$

$$x^2 - 4x = -x + 4$$

Notes:

$$x^2 - 3x - 4 = 0$$
$$(x - 4)(x + 1) = 0$$
$$x = -1, 4$$

Now that we have our x coordinates, we need to find the corresponding y coordinates. We'll use the first equation, since it's a bit simpler to work with. Substituting in $x = 4$ gives us $0 = y^2$, or $y = 0$. Substituting in $x = -1$ gives $-5 = y^2$. Here, this gives us an imaginary answer for y, so we do not get an additional intersection point. The only intersection point for these equations is

$$\boxed{(4, 0)}$$

Equating the Functions

The next method we will discuss works well when both functions are explicit, or are given in function notation. For this method, we will first solve each equation for the same variable, set the two equal to each, and solve.

Example 71   **Points of Intersection: Equating**
Find all points of intersection of $f(x) = x^2 + 1$ and $g(x) = x + 1$

    Solution   Here, both equations are given using function notation; this means that really $f(x)$ tells us the value of the y coordinate at x, so we can replace it with y: $y = x^2 + 1$. Similarly, $g(x)$ tells us the value of the y coordinate at x for the other function: $y = x + 1$. Since y is isolated in both, we will set the two equal to each other and solve for x:

$$x^2 + 1 = x + 1$$
$$x^2 - x = 0$$
$$x(x - 1) = 0$$
$$x = 0, 1$$

Now, we just need to find the y coordinates. We can use either $f(x)$ or $g(x)$ to do this; $g(x)$ is simpler so we will use it. We get that $g(0) = 1$ and $g(1) = 2$. Therefore, we have two points of intersection:

$$\boxed{(0, 1) \text{ and } (1, 2)}$$

Notes:

Elimination

The third method we will discuss is a bit different than the other methods we have seen. This method also requires strong algebra skills. The main advantage of this method won't be obvious until the next section of this book, because it is the most useful when we have a system of two or more linear equations. Here, we will only show how to use it with two variables, but the idea extends nicely (this means that it is easy to adapt this method to other more complicated situations). For elimination, we will take each equation, multiply the entire equation by a constant, and add the equations together in such a way that one variable is eliminated.

Example 72   **Points of Intersection: Elimination**
Find all points of intersection of $2x + 3y = 2$ and $-x + y = 4$.

Solution   We will first try to eliminate x from both equations. The first equation has $2x$ and the second has $-x$. If we multiply the second equation by 2 and add it to the first, the x terms will cancel out:

$$2x + 3y = 2$$
$$+2(-x + y = 4)$$

or:

$$2x + 3y = 2$$
$$+(-2x + 2y = 8)$$
$$\overline{5y = 10}$$

Notice that we lined up our variables and treated this like a big addition problem. Keeping the variables lined up makes our work easier to follow.

Now, we can take the result and easily solve for y, getting $y = 2$. We can now use y to find x. Either equation will work, but we will use the second one: $-x + (2) = 4$, or $x = -2$. This gives use one point of intersection at

$$\boxed{(-2, 2)}$$

Elimination is a particularly flexible method. To illustrate this, we will solve the problem again, but this time we will eliminate y first.

Example 73   **Points of Intersection: Elimination**
Find all points of intersection of $2x + 3y = 2$ and $-x + y = 4$.

Solution   The first equation has $3y$ and the second has y. We will multiply the first equation by $-\frac{1}{3}$ and add it to the second equation:

Notes:

$$-\tfrac{1}{3}(\ 2x\ +3y= 2\ )$$
$$+\quad (\ -x + y = 4\ )$$
$$-\tfrac{5}{3}x \qquad = \tfrac{10}{3}$$

Solving $-\tfrac{5}{3}x = \tfrac{10}{3}$ gives us $x = -2$, and substituting into either equation gives us $y = 2$. We get the same intersection point:

$$(-2, 2)$$

Additionally, we could have multiplied the second equation, $-x + y = 4$, by 3 and subtracted from the first to eliminate y first. With elimination, it is best to do a little planning to figure out what variable will be easiest to eliminate first, and what combinations will keep the numbers simple.

Graphing

Now, we will briefly discuss a common method used by students: graphing. While graphing is a great way to help determine how many points of intersection exist and the approximate coordinates, it will not give you an exact set of coordinates, unless you use a calculator or computer. In calculus, having the exact values is necessary. In example 69, we ended up with two points of intersection: $(\tfrac{1+\sqrt{7}}{4}, \tfrac{-1+\sqrt{7}}{2})$ and $(\tfrac{1-\sqrt{7}}{4}, \tfrac{-1-\sqrt{7}}{2})$. If we had graphed to find these points, we would not have found the exact coordinates, and at best would have ended up with approximations. For this reason, we do not recommend relying solely on graphing for finding points of intersection. Sometimes the coordinates will be integers and the graph will be easy to read, but as in this example, often it is impossible to get the answer you need from the graph.

Notes:

Exercises 3.2

Terms and Concepts

1. In which situations is substitution a more appropriate solution method than equating the functions?

2. Is $y = x^3 + 5x - 7$ an implicitly or explicitly defined function? Explain.

3. Is $xy + y^2 - y = 2x + 6$ an implicitly or explicitly defined function? Explain.

4. Describe the pros and cons of using graphing to find the point(s) of intersection.

Problems

In exercises 5 – 8, determine the maximum possible number of intersections for the described functions.

5. Two linear functions with different slopes

6. A linear function and a quadratic function

7. Two explicitly defined quadratic functions

8. A cubic function and a constant function

In exercises 9 – 12, determine the minimum possible number of intersections for the described functions.

9. Two linear functions with different slopes

10. A linear function and a quadratic function

11. Two explicitly defined quadratic functions

12. A cubic function and a constant function

In exercises 13 – 18, find all points of intersection between the given functions.

13. $y = x^2 - 1$ and $y = x - 1$

14. $x^2 + y^2 = 1$ and $4y = 3x$

15. $y - 1 = \sqrt{3x}$ and $y = x + 1$

16. $y = x^2 - 3x + 2$ and the x-axis

17. $y = x^2 - 3x + 2$ and $y = 5$

18. $y + 2x = 5$ and $y + 3 = x^3 - 7x^2 + 12x$

In exercises 19 – 22, sketch the region bounded by the given functions and determine all intersection points.

19. $y = x^2$ and $y = x$

20. $y = x^2$ and $y = x + 2$

21. $y = x^2$ and $y = \sqrt{x}$

22. $3y + 2x = 6$, the x-axis, and the y-axis (hint: sketch before looking for the intersection points)

3.3   Fractions and Partial Fractions Decomposition

In calculus, you will run into many situations where you need to simplify fractions; in differential calculus, when you take a derivative of a quotient of two functions, the result will be an even more complicated quotient that will require simplification. Additionally, when working with rational functions (functions that are a quotient of two polynomials), simplifying can help identify key features of the function. In integral calculus and when working with inverse Laplace transforms in differential equation, you will need to take a fraction and split it into several simpler fractions through a process called partial fraction decomposition. In this section, we will discuss many of the skills you will need when working with fractions in calculus.

Simplifying Fractions

When mathematicians talk about simplifying fractions they can be referring to combining fractions that are being added into a single fraction, removing any common factors from the numerator and denominator, and/or rewriting fractions that have nested fractions in the numerator or denominator. First, we'll discuss how to combine multiple fractions.

When adding or subtracting any fractions, the first step is to get a common denominator. This builds off of the ideas we learn about fractions as a child; the denominator tells us how many pieces we split the item into and the numerator tells us how many pieces we are using. For example, $\frac{2}{3}$ means we split the item into 3 pieces and are using 2 of them. Before we can combine fractions, we need to make sure all of the pieces are the same size by having the same denominator. As we saw in Section 1.1, we can make sure that are denominators are the same by multiplying by 1 in a sneaky way; for example if we want to add $\frac{2}{3}$ and $\frac{1}{8}$, we can multiply by $\frac{8}{8}$ and by $\frac{3}{3}$ respectively. Since we are multiplying by the "missing" factor for each, both will have the same denominator: $\frac{16}{24}$ and $\frac{3}{24}$. We can do the same even when our fraction contains variables.

Example 74　　**Combining Fractions**

Simplify $\dfrac{3}{x+2} - \dfrac{x+1}{x-2}$.

Solution　　First, we will multiply by the missing factors. We will multiply the first term by $\dfrac{x-2}{x-2}$ and the second term by $\dfrac{x+2}{x+2}$. This gives us:

Notes:

$$\frac{3}{x+2} - \frac{x+1}{x-2} = \frac{3}{x+2} \times \frac{x-2}{x-2} - \frac{x+1}{x-2} \times \frac{x+2}{x+2}$$

$$= \frac{3(x-2)}{(x+2)(x-2)} - \frac{(x+1)(x+2)}{(x-2)(x+2)}$$

$$= \frac{3x-6}{x^2-4} - \frac{x^2+3x+2}{x^2-4}$$

Notice that when we multiplied we were careful to include parentheses since we know that we have implied parentheses when we work with fractions. Now that both fractions have the same denominator, we can combine them. We must do this carefully; we are subtracting so we will need to distribute the negative correctly.

$$\frac{3x-6}{x^2-4} - \frac{x^2+3x+2}{x^2-4} = \frac{3x-6-(x^2+3x+2)}{x^2-4}$$

$$= \frac{-x^2-8}{x^2-4}$$

Our final answer is

$$\boxed{\frac{3}{x+2} - \frac{x+1}{x-2} = \frac{-x^2-8}{x^2-4}}$$

Simplifying a fraction can also mean that we are looking for common factors of the numerator and the denominator. If we examine the result from our previous example, we see that the denominator can be factored: $x^2 - 4 = (x+2)(x-2)$. However, the numerator is irreducible. This means that it has no linear factors, so the numerator and denominator have no factors in common and cannot be simplified any further.

This type of simplifying can be confusing for students; it's really tempting to see a fraction like $\frac{x^2+1}{x+1}$ and to "simplify" it by crossing out the ones. However, this is not correct because it ignores the implied parentheses: $\frac{(x^2+1)}{(x+1)}$. The one is tied to the rest of the terms and cannot be separated in this manner. This becomes a bit clearer if you try substituting in a number for x, such as $x = 2$. Let's look at an example where we do have common factors.

Notes:

Example 75   Simplifying a Fraction

Simplify $\dfrac{2x^3 + 10x^2 + 12x}{2x^3 - 8x}$.

Solution   The first step here is to factor both the numerator and the denominator. We won't show those steps here, but you should verify our result. Once we have factored both, we will see if we have any common factors; if we do we can remove them from both the numerator and the denominator.

$$\frac{2x^3 + 10x^2 + 12x}{2x^3 - 8x} = \frac{2x(x + 3)(x + 2)}{2x(x + 2)(x - 2)}$$
$$= \frac{x + 3}{x - 2}, x \neq 0, -2$$

We see that both the numerator and the denominator have $2x$ and $x + 2$ as factors; this means we can eliminate these terms. Notice that we have to add a domain restriction. The original form of the fraction is not defined at $x = 0$ or $x = 2$ since both values make the denominator 0. In order to truly have the same meaning as the original function, we need to note that we cannot use these values of x. This is why we have the additional note of $x \neq 0, -2$ as part of our solution. There are no other common factors, so our final answer is

$$\boxed{\frac{2x^3 + 10x^2 + 12x}{2x^3 - 8x} = \frac{x + 3}{x - 2}, x \neq 0, -2}$$

Now, let's take a look at simplifying when we have complex fractions. Here, complex does not mean that we are working with imaginary numbers, rather that we have a fraction nested inside of a fraction. When we have complex fractions, the first step is to make sure the entire numerator is as simplified as possible and that the entire denominator is as simplified as possible. We'll work three different examples that already have simplified numerators and simplified denominators, but do not neglect this first step as it is critical in working these problems correctly.

Example 76   Complex Fractions

Simplify $\dfrac{x + 1}{\frac{x-1}{x^2}}$.

Solution   Again, note that both the numerator and denominator are as simplified as possible. Here, the nested fraction is in the denominator. When

Notes:

dividing by a fraction, we can instead multiply by the reciprocal (think about dividing a number by $\frac{1}{2}$; it is equivalent to multiplying by $\frac{2}{1}$).

$$\frac{x+1}{\frac{x-1}{x^2}} = (x+1) \div \frac{x-1}{x^2}$$

$$= (x+1) \times \frac{x^2}{x-1}$$

$$= \frac{(x+1)(x^2)}{x-1}$$

$$= \frac{x^3 + x^2}{x-1}$$

There are no common factors, so we are done, and our final answer is

$$\boxed{\frac{x+1}{\frac{x-1}{x^2}} = \frac{x^3 + x^2}{x-1}}$$

Example 77   **Complex Fractions**

Simplify $\dfrac{\frac{2}{x+1}}{x+2}$.

   Solution   Here, the nested fraction is in the numerator. For this case, we can simply rewrite a little bit; instead of dividing by $x+2$, we can multiply by $\frac{1}{x+2}$. This is analogous to multiplying by one half instead of dividing by 2; both have the same meaning.

$$\frac{\frac{2}{x+1}}{x+2} = \frac{2}{x+1} \div (x+2)$$

$$= \frac{2}{x+1} \times \frac{1}{x+2}$$

$$= \frac{2(1)}{(x+1)(x+2)}$$

$$= \frac{2}{x^2 + 3x + 2}$$

Notes:

There are no common factors, so we are done.

$$\frac{\frac{2}{x+1}}{x+2} = \frac{2}{x^2+3x+2}$$

Example 78   **Complex Fractions**

Simplify $\dfrac{\frac{x}{x+1}}{\frac{x-2}{x-1}}$.

Solution   Here we will use the ideas from both of the previous examples. We will multiply the numerator by the reciprocal of the denominator:

$$\frac{\frac{x}{x+1}}{\frac{x-2}{x-1}} = \frac{x}{x+1} \div \frac{x-2}{x-1}$$

$$= \frac{x}{x+1} \times \frac{x-1}{x-2}$$

$$= \frac{(x)(x+1)}{(x-1)(x-2)}$$

$$= \frac{x^2+1}{x^2-3x+2}$$

There are no common factors, so we are done.

$$\frac{\frac{x}{x+1}}{\frac{x-2}{x-1}} = \frac{x^2+1}{x^2-3x+2}$$

Partial Fraction Decomposition

Our next topic is partial fraction decomposition. With partial fraction decomposition, our goal is to take a fraction with a polynomial numerator and a polynomial denominator and write it as the sum of several fractions that have simpler denominators. For example, we can write $\frac{6x+16}{x^2+5x+6}$ as $\frac{2}{x+3} + \frac{4}{x+2}$ (this

Notes:

is a good place to practice your fraction combining skills by verifying that these are equal). In many situations, particularly when performing integration, this second form is much easier to work with. To do this, our first step is to factor the denominator. When we factor, we will end up with linear factors and/or irreducible quadratic factors. These factors may only appear once, or may be repeated (for example, for $x^2 + 2x + 1$, we say $x + 1$ is repeated twice since $x^2 + 2x + 1 = (x + 1)^2$).

With our decomposition, we want to write the original fraction as the sum of many fractions; we will need one fraction for each factor. If a factor is repeated, it will need one fraction for each time it is repeated. The factors will be the denominators of the new fractions. Remember, the factors used to make the new fraction denominators must combine, through multiplication, to give us the original denominator. For linear factors, the numerator will be a constant and for quadratic factors the numerator will be linear. Once we have determined how we are splitting up ("decomposing") our original fraction, we will use our algebra skills to determine exactly what the numerators look like. Let's look at some examples; in all of our examples the denominator will already be factored; in practice you will often need to do the factorization as your first step.

Example 79   Partial Fraction Decomposition: Linear Factors

Perform a partial fraction decomposition on $\dfrac{3}{(x + 1)(x - 2)}$.

Solution   Since the denominator has two factors, we will be decomposing into two fractions. Each term is linear, so each of our new fractions will have a constant numerator. We'll use A and B as the numerators for now, and we will solve for these two values later. So far, we have:

$$\frac{3}{(x + 1)(x - 2)} = \frac{A}{x + 1} + \frac{B}{x - 2} \tag{3.2}$$

It does not matter which fraction comes first, nor does it matter what letters we use in the numerators, so long as we don't use the same letter twice. Our next step is to determine the appropriate values for A and B. To make this easier, we will multiply everything in equation 3.2 by $(x + 1)$ and by $(x - 2)$. This will eliminate all of the fractions.

$$(x + 1)(x - 2)\left[\frac{3}{(x + 1)(x - 2)}\right] = (x + 1)(x - 2)\left[\frac{A}{x + 1} + \frac{B}{x - 2}\right]$$

$$\frac{3(x + 1)(x - 2)}{(x + 1)(x - 2)} = \frac{A(x + 1)(x - 2)}{x + 1} + \frac{B(x + 1)(x - 2)}{x - 2}$$

$$3 = A(x - 2) + B(x + 1)$$

Notes:

You might be tempted to distribute on the right side, but it will be easier to solve for A and B if we don't. If we have the right values of A and B, this last statement, $3 = A(x-2) + B(x+1)$, is true for all values of x. We will exploit this. Right now, we are multiplying A by $(x-2)$. We can make this A term disappear if we substitute in $x = 2$. When we do this, we get:

$$3 = 0 + B(2+1)$$
$$3 = 3B$$
$$B = 1$$

We can use a similar technique by substituting in $x = -1$ and making B disappear:

$$3 = A(-1-2) + 0$$
$$3 = A(-3)$$
$$A = -1$$

Now that we have the values of A and B, we can complete the decomposition:

$$\boxed{\frac{3}{(x+1)(x-2)} = \frac{-1}{x+1} + \frac{1}{x-2}}$$

With partial fraction decomposition, the order of the times is up to you. We could have started out this problem with

$$\frac{3}{(x+1)(x-2)} = \frac{A}{x-2} + \frac{B}{x+1}$$

instead of

$$\frac{3}{(x+1)(x-2)} = \frac{A}{x+1} + \frac{B}{x-2}$$

The values for A and B would be different, but the final answer would be the same.

As we noted above, we may have repeated factors in our denominator, and when we do we will need a separate fraction for each time it is repeated. These fractions will all have this repeated factor in the denominator, but raised to a higher power each time: in the first fraction we will just have the factor, in the second fraction we will have the factor squared, in the third we will have the factored cubed, etc. Let's take a look at an example.

Notes:

Example 80 **Partial Fraction Decomposition: Repeated Factors**

Perform a partial fraction decomposition on $\dfrac{5x^3 + 16x^2 + 16x + 6}{(x+2)(x+1)^3}$.

Solution Here, the denominator is already factored for us, so the first step is already complete. We see that we have one factor that is only repeated once, $x + 2$, and another factor that is repeated three times, $x + 1$. This means we will decompose into four fractions, with denominators of $x+2$, $x+1$, $(x+1)^2$, and $(x+1)^3$. Since both factors are linear, each fraction will have a constant in the numerator. So, the decomposition will look like:

$$\frac{5x^3 + 16x^2 + 16x + 6}{(x+2)(x+1)^3} = \frac{A}{x+2} + \frac{B}{x+1} + \frac{C}{(x+1)^2} + \frac{D}{(x+1)^3}$$

As in the previous example, we will multiply both sides by the denominator of the original fraction, $(x+2)(x+1)^3$. This will eliminate the fractions.

$$(x+2)(x+1)^3\left[\frac{5x^3 + 16x^2 + 16x + 6}{(x+2)(x+1)^3}\right] =$$

$$= (x+2)(x+1)^3\left[\frac{A}{x+2} + \frac{B}{x+1} + \frac{C}{(x+1)^2} + \frac{D}{(x+1)^3}\right]$$

Then,

$$\frac{(5x^3 + 16x^2 + 16x + 6)(x+2)(x+1)^3}{(x+2)(x+1)^3} =$$

$$= \frac{A(x+2)(x+1)^3}{x+2} + \frac{B(x+2)(x+1)^3}{x+1} + \frac{C(x+2)(x+1)^3}{(x+1)^2} + \frac{D(x+2)(x+1)^3}{(x+1)^3}$$

Finally,

$$5x^3 + 16x^2 + 16x + 6 = A(x+1)^3 + B(x+2)(x+1)^2 + C(x+2)(x+1) + D(x+2)$$

Now, we'll use the same method we used in the previous example; by choosing appropriate values of x to substitute into our equation, we will be able to eliminate terms. Every term except for the A term is being multiplied by $(x+2)$,

Notes:

so if we substitute $x = -2$, the B, C, and D terms will all become zero:

$$5(-2)^3 + 16(-2)^2 + 16(-2) + 6 = A(-2 + 1)$$
$$5(-8) + 16(4) + 16(-2) + 6 = A(-1)$$
$$-40 + 64 - 32 + 6 = -A$$
$$-2 = -A$$
$$A = 2$$

Next, by substituting $x = -1$, we can find the value for D:

$$5(-1)^2 + 16(-1)^2 + 16(-1) + 6 = D(-1 + 2)$$
$$5(-1) + 16(1) + 16(-1) + 6 = D(1)$$
$$-5 + 16 - 16 + 6 = D$$
$$1 = D$$

We now have values for A and for D, but unfortunately our method will not work to help us find B and C since each of these terms is multiplied by both factors. We'll take a similar approach, however. We noted that these statements are true for all values of x, so we can choose some easy-to-work-with values to substitute in. Currently, we have

$$5x^3 + 16x^2 + 16x + 6 = 2(x+1)^3 + B(x+2)(x+1)^2 + C(x+2)(x+1) + (x+2)$$

We'll start by substituting in $x = 0$ since this keeps the arithmetic easy. We get

$$5(0)^3 + 16(0)^2 + 16(0) + 6 = 2(0+1)^3 + B(0+2)(0+1)^2 + C(0+2)(0+1) + (0+2)$$
$$6 = 2(1)^3 + B(2)(1)^2 + C(2)(1) + 2$$
$$6 = 2 + 2B + 2C + 2$$
$$2 = 2B + 2C$$
$$1 = B + C$$

This doesn't give us enough information to find values for B and C, so we will need another equation. To get this equation, we will substitute $x = 1$:

$$5(1)^3 + 16(1)^2 + 16(1) + 6 = 2(1+1)^3 + B(1+2)(1+1)^2 + C(1+2)(1+1) + (1+2)$$
$$5 + 16 + 16 + 6 = 2(2^3) + B(3)(2)^2 + C(3)(2) + 3$$
$$43 = 16 + 12B + 6C + 3$$
$$24 = 12B + 6C$$
$$4 = 2B + C$$

Now, we have two equations: $1 = B + C$ and $4 = 2B + C$. We can now use the methods we learned when finding points of intersection; we have two

Notes:

equations with 2 values that we need to find. We will use elimination to solve since both equations have C with the same coefficient, but you can use any of the methods we learned. We will subtract $1 = B + C$ from $4 = 2B + C$ to get $3 = B$. We can substitute $B = 3$ into $1 = B + C$ and solve for C to get $C = -2$. Finally, we have:

$$\frac{5x^3 + 16x^2 + 16x + 6}{(x+2)(x+1)^3} = \frac{2}{x+2} + \frac{3}{x+1} + \frac{-2}{(x+1)^2} + \frac{1}{(x+1)^3}$$

As you can see, partial fraction decomposition can be a tedious process. The biggest issues people encounter when performing a partial fraction decomposition are algebra/arithmetic mistakes and copy errors. These errors tend to be caused by rushing; with a process like partial fractions, it is better to work slowly, carefully, and methodically to avoid these errors, lest you have to start over from the beginning.

We're not quite done with partial fraction decompositions yet. We've covered how to deal with linear factors, even with repetitions, but we haven't yet seen how to work with irreducible quadratic factors. As with linear factors, we will decompose into one fraction per factor. The difference is in the numerator. For the irreducible quadratic factors, the numerators need to be linear. Let's take a look:

Example 81   **Partial Fraction Decomposition: Quadratic Factors**

Perform a partial fraction decomposition on $\dfrac{5x^2 - x + 2}{(x-1)(x^2+1)}$.

Solution   Let's dive right in and start our decomposition. We have two factors, so we will decompose into two fractions:

$$\frac{5x^2 - x + 2}{(x-1)(x^2+1)} = \frac{A}{x-1} + \frac{Bx+C}{x^2+1}$$

As always, the linear factor gets a constant in the numerator. The quadratic factors get a linear numerator. We'll multiply by the original fraction's denominator to eliminate the fractions:

$$(x-1)(x^2+1)\left[\frac{5x^2 - x + 2}{(x-1)(x^2+1)}\right] = (x-1)(x^2+1)\left[\frac{A}{x-1} + \frac{Bx+C}{x^2+1}\right]$$

Notes:

$$\frac{(5x^2 - x + 2)(x - 1)(x^2 + 1)}{(x - 1)(x^2 + 1)} = \frac{A(x - 1)(x^2 + 1)}{x - 1} + \frac{(Bx + C)(x - 1)(x^2 + 1)}{x^2 + 1}$$

$$5x^2 - x + 2 = A(x^2 + 1) + (Bx + C)(x - 1)$$

Notice the parentheses around $Bx + C$ in the last line; without these parentheses we would not be multiplying correctly. We'll start off with our favorite method and substitute $x = 1$ to eliminate the $Bx + C$ term:

$$5(1)^2 - (1) + 2 = A((1)^2 + 1) + (B(1) + C)(1 - 1)$$
$$5(1) - 1 + 2 = A(1 + 1) + 0$$
$$6 = 2A$$
$$A = 3$$

Finding B and C will be fairly quick. We already have a value for A, and if we substitute $x = 0$, we can eliminate B (since it is multiplied by x). Let's take a look:

$$5(0)^2 - (0) + 2 = A(0^2 + 1) + (B(0) + C)(0 - 1)$$
$$2 = A(1) + (C)(-1)$$
$$2 = A - C$$

Since we know $A = 3$, we get $C = 1$. Now that we know A and C, we can substitute in a third value for x to find B, or we can simplify both sides to find B. We'll show this second method.

$$5x^2 - x + 2 = A(x^2 + 1) + (Bx + C)(x - 1)$$
$$5x^2 - x + 2 = (3)(x^2 + 1) + (Bx + (1))(x - 1)$$
$$5x^2 - x + 2 = 3x^2 + 3 + Bx^2 - Bx + x - 1$$
$$5x^2 - x + 2 = (3 + B)x^2 + (-B + 1)x + 2$$

Using the x^2 terms, we have $5x^2 = (3 + B)x^2$, so $5 = 3 + B$, or $B = 2$. We get the same result if we match the x terms. Altogether, we have:

$$\boxed{\frac{5x^2 - x + 2}{(x - 1)(x^2 + 1)} = \frac{3}{x - 1} + \frac{2x + 1}{x^2 + 1}}$$

Lastly, we could have a fraction with repeated irreducible quadratics. We won't show the full solution for one of these, but we will show the initial setup.

Notes:

Just like with repeated linear factors we will need a fraction for each time the quadratic is repeated, and just like the previous example, they will each have a linear numerator. For example, we would have the following decomposition:

$$\frac{2x + 6}{(x^2 + 4)^2(x + 1)} = \frac{A}{x + 1} + \frac{Bx + C}{x^2 + 4} + \frac{Dx + E}{(x^2 + 4)^2}$$

We would then solve for A, B, C, D, and E using the same methods we have used in our other examples.

Notes:

Exercises 3.3

Terms and Concepts

1. Can the fraction $\dfrac{x+2}{x^2+2}$ be simplified? Explain.

2. In the fraction $\dfrac{2}{(x+3)^2(x+2)}$ are there any repeated factors? If so, what factor(s) are repeated, and how many times?

3. What is meant by an irreducible quadratic?

4. Give an example of an irreducible quadratic.

Problems

Simplify the given expression in exercises 5 – 9.

5. $\dfrac{5}{18} - \dfrac{5}{12}$

6. $\dfrac{x}{b} - \dfrac{b}{x}$

7. $\dfrac{x}{y^2} - \dfrac{x}{x+y}$

8. $\dfrac{\frac{1}{x} - \frac{x+2}{x^2}}{\frac{4}{x^2} - \frac{x^2+1}{x^3}}$

9. $\dfrac{\frac{1}{x-b} - \frac{1}{x}}{b}$

In exercises 10 – 16, decompose the given fraction. Do not solve for A, B, etc.

10. $\dfrac{x-8}{(x+2)^3}$

11. $\dfrac{4}{(s-1)^2(2s-5)(s+3)}$

12. $\dfrac{5t^2+11t-9}{(t+1)^3(t^2+1)^2}$

13. $\dfrac{6x}{(x-4)(x^2+x+5)}$

14. $\dfrac{3x-7}{x^4-1}$

15. $\dfrac{2s}{s^3+1}$

16. $\dfrac{11}{t^2-6t+5}$

In exercises 17 – 22, fully decompose the given fraction.

17. $\dfrac{x+5}{x^2+x-2}$

18. $\dfrac{1}{x^2-a^2}$

19. $\dfrac{2s^2-s+4}{s^3+4s}$

20. $\dfrac{y-1}{y^2+3y+2}$

21. $\dfrac{4x}{x^3-x^2-x+1}$

22. $\dfrac{x^2+2x-1}{2x^3+3x^2-2x}$

3.4   Introduction to Trigonometric Functions

In this section, we will introduce trigonometric functions. We will examine relationships between these functions and will discuss how to evaluate these functions for commonly used inputs. Trigonometric functions are used frequently in calculus and later courses due to the wide range of phenomena that can be modeled by these functions, everything from airflow in our bronchial tubes to earthquake vibrations in a building.

Trigonometric Definitions

In this section, we will focus on six trigonometric functions: sine, cosine, tangent, cosecant, secant, and cotangent. These functions are all related to each other, and typically mathematicians focus on sine and cosine since the other four functions can all be expressed in terms of sine and cosine. These relationships are:

- $\tan(\theta) = \dfrac{\sin(\theta)}{\cos(\theta)}$

- $\csc(\theta) = \dfrac{1}{\sin(\theta)}$

- $\sec(\theta) = \dfrac{1}{\cos(\theta)}$

- $\cot(\theta) = \dfrac{\cos(\theta)}{\sin(\theta)}$

Here $\tan(\theta)$ is the notation commonly used for the tangent function; $\csc(\theta)$ is the notation for the cosecant function; $\sec(\theta)$ is the notation for the secant function; and $\cot(\theta)$ is the notation for the cotangent function. Notice that each of these take an input value; by itself "sin " has no more meaning than "$\sqrt{\ }$" has; they are all functions and all require an input. Additionally, notice that we use θ (the Greek letter theta) as our input variable. Mathematicians typically (but not always) use Greek letters when referring to angles, so you will often see θ and other Greek letters used to label angles, like α (alpha) and β (beta).

The Unit Circle

All trigonometric functions have repeating patterns; as such, most mathematicians use a tool known as the unit circle to help evaluate these functions. The unit circle is a circle of radius 1 where we will label key points. Before we see the unit circle, let's talk a bit about how to use the unit circle.

Notes:

For each trigonometric function, mathematicians typically think of the input as measuring an angle. You probably are used to measuring angles in degrees, but in calculus we will measure angles with a unit called *radians*. Radians are just a different type of unit for measuring angles, just like feet and meters are different units for measuring distance. You are probably familiar with the idea that a complete circle has $360°$; in radians this is the same as 2π radians. Mathematicians prefer radians to degrees for several reasons. First, is that radians relate angles to arc length, the distance around the circle. For a full circle, we have a special name for the arc length: circumference. You may remember that the formula for circumference is $C = 2\pi r$; with radians this is the same as saying the size of the angle times the radius. A second reason will show itself in calculus when you learn derivatives.

When using the unit circle, we measure the angle as we move counter clockwise, using the positive x-axis as our starting point. This means that by the time we reach the positive y-axis we have gone a quarter of the way around the circle; we have swept over an angle of $\frac{2\pi}{4} = \frac{\pi}{2}$ radians. This is illustrated in Figure 3.1. If we move around the circle in the opposite direction, clockwise, we say the angles are negative. This means that we have swept over $-\frac{3\pi}{2}$ radians if we move to the positive y-axis in the clockwise direction.

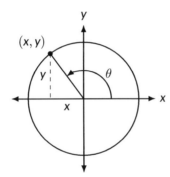

Figure 3.1: Measuring angles; reproduced from A^B_EX Calculus, Version 3.0

Once we have swept over the angle we are interested in, we will need to know the x and y coordinates of the associated point on the circle. The x coordinate at the point is the value of $\cos(\theta)$ for that angle and the y coordinate is the value of $\sin(\theta)$ for that angle. The unit circle, shown in Figure 3.2, shows these coordinates for a variety of common inputs. Here, the unit circle is shown with both angles measured in both degrees and radians, but remember that we are focusing on radians.

Notes:

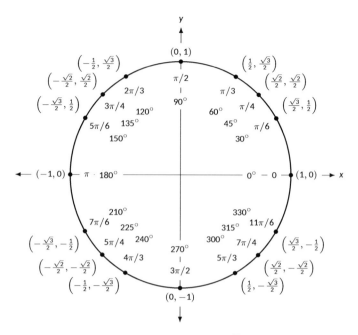

Figure 3.2: The unit circle; reproduced from A$\overset{P}{E}$X Calculus, Version 3.0

Let's look at a few examples of how to use the unit circle to evaluate trigonometric functions. Remember, the unit circle will help us determine which x and y coordinates pair with each angle. The x coordinate gives us the value of cosine for the angle and the y coordinate gives us the value of sine for the angle.

Example 82   **Evaluating Trigonometric Functions**
Evaluate each of the following:

1. $\sin\left(\frac{\pi}{4}\right)$

2. $\tan\left(\frac{\pi}{6}\right)$

3. $\sec\left(3\pi\right)$

4. $\cos\left(-\frac{3\pi}{4}\right)$

5. $\csc\left(\frac{2\pi}{3}\right)$

6. $\cot\left(\frac{5\pi}{6}\right)$

Solution   Let's get started:

1. Here, our input angle is $\frac{\pi}{4}$. This corresponds to the coordinate pair $\left(\frac{\sqrt{2}}{2}, \frac{\sqrt{2}}{2}\right)$. We are interested in the value of sine, so we want to look at the y coordinate. This gives us

Notes:

$$\sin\left(\frac{\pi}{4}\right) = \frac{\sqrt{2}}{2}$$

2. Here, our input angle is $\frac{\pi}{6}$ which corresponds to the coordinate pair $\left(\frac{\sqrt{3}}{2}, \frac{1}{2}\right)$. The coordinate pairs give us the values for cosine and sine, but doesn't directly give us a value for tangent, so we will need to use our definition of tangent. We have

$$\tan\left(\frac{\pi}{6}\right) = \frac{\sin\left(\frac{\pi}{6}\right)}{\cos\left(\frac{\pi}{6}\right)}$$

$$= \frac{\frac{\sqrt{3}}{2}}{\frac{1}{2}}$$

$$= \frac{\sqrt{3}}{2} \times \frac{2}{1}$$

$$= \sqrt{3}$$

So, our final answer is

$$\tan\left(\frac{\pi}{6}\right) = \sqrt{3}$$

3. Here, our input angle is 3π. There is no angle labeled as 3π on the unit circle, so this one requires a bit more thought. We said earlier that the circle has 2π radians, so if we completely go around the circle, we have covered 2π. We need to go another $3\pi - 2\pi = \pi$ radians, so we can use the coordinates at π radians to get our values for 3π radians. At π radians, our coordinates are $(-1, 0)$. This gives us:

$$\sec\left(3\pi\right) = \frac{1}{\cos\left(3\pi\right)}$$

$$= \frac{1}{-1}$$

$$= -1$$

So, we have

Notes:

$$\sec(3\pi) = -1$$

4. Here, our input angle is $-\frac{3\pi}{4}$. This means we are covering the unit circle by moving in a clockwise direction. We need to start at 0 radians, and move $\frac{3\pi}{4}$ radians clockwise. This would put us at $\frac{5\pi}{4}$ radians. The coordinates are $\left(-\frac{\sqrt{2}}{2}, -\frac{\sqrt{2}}{2}\right)$, so

$$\cos\left(-\frac{3\pi}{4}\right) = -\frac{\sqrt{2}}{2}$$

5. Here, our input angle is $\frac{2\pi}{3}$ radians, which gives coordinates of $\left(-\frac{1}{2}, \frac{\sqrt{3}}{2}\right)$. Cosecant relies on sine, so we have

$$\csc\left(\frac{2\pi}{3}\right) = \frac{1}{\sin\left(\frac{2\pi}{3}\right)}$$
$$= \frac{1}{\frac{\sqrt{3}}{2}}$$
$$= \frac{2}{\sqrt{3}}$$
$$= \frac{2}{\sqrt{3}} \times \frac{\sqrt{3}}{\sqrt{3}}$$
$$= \frac{2\sqrt{3}}{3}$$

Notice that we rationalized the denominator. This means that we rewrote the fraction so that it would not have an irrational number, $\sqrt{3}$, in the denominator. This is a common last step in mathematics to standardize the form of the answer. Most answer keys will write the answer in the rationalized form, so it is a good habit to rationalize the denominator so that you can check your answers. Our final answer is

$$\csc\left(\frac{2\pi}{3}\right) = \frac{2\sqrt{3}}{3}$$

Notes:

6. Here, we have an angle of $\frac{5\pi}{6}$, which has coordinates $\left(-\frac{\sqrt{3}}{2}, \frac{1}{2}\right)$. This gives us

$$
\begin{aligned}
\cot\left(\frac{5\pi}{6}\right) &= \frac{\cos\left(\frac{5\pi}{6}\right)}{\sin\left(\frac{5\pi}{6}\right)} \\
&= \frac{-\frac{\sqrt{3}}{2}}{\frac{1}{2}} \\
&= \frac{-\sqrt{3}}{2} \times \frac{2}{1} \\
&= -\sqrt{3}
\end{aligned}
$$

$$
\boxed{\cot\left(\frac{5\pi}{6}\right) = -\sqrt{3}}
$$

Properties of Trigonometric Functions

Notice that on the unit circle, the values of the x and y coordinates range between -1 and 1. Since these are the output values of the sine and cosine functions, we say that the *range*, or the output values, of sine and cosine is $[-1, 1]$, meaning that the output can get as small as -1 and as large as 1. Additionally, sine and cosine are defined for any input value, so for each the domain is $(-\infty, \infty)$.

The other four trigonometric functions are all defined in terms of sine and cosine, where either sine or cosine is in the denominator of a fraction. This means that the domains of these functions are limited. Both secant and tangent have cosine in the denominator, meaning that they will be undefined anytime cosine is 0. Cosine is 0 at the odd multiples of $\frac{\pi}{2}$. This tells us that these odd multiple of $\frac{\pi}{2}$, values like $-\frac{3\pi}{2}$, $-\frac{\pi}{2}$, $\frac{\pi}{2}$, and $\frac{3\pi}{2}$ are not part of the domain for tangent or for secant. Like sine and cosine, the output values of secant is limited; its range is $(-\infty, -1] \cup [1, \infty)$, but the range for tangent is not limited; its range is $(-\infty, \infty)$.

Cosecant and cotangent both have sine in the denominator for their definitions; this means they are undefined whenever sine is 0. Sine is 0 at the integer multiples of π: $-5\pi, -3\pi, -\pi, \pi, 3\pi$, etc. As with secant, the output values of cosecant are limited so its range is also $(-\infty, -1] \cup [1, \infty)$; the range of cotangent is not limited and is $(-\infty, \infty)$.

Notes:

Exercises 3.4

Terms and Concepts

1. Explain why the domain of tangent is restricted.

2. Explain why the domain of cosecant is restricted.

3. Explain what is meant by the range of a function.

4. What do the coordinates on the unit circle tell you?

5. Sketch the unit circle from memory. Use Figure 3.2 to check your work and add in any values you could not remember.

Problems

Evaluate each statement given in exercises 6 – 10.

6. $\tan\left(\dfrac{\pi}{4}\right)$

7. $\cos\left(\dfrac{-\pi}{4}\right)$

8. $\sin\left(\dfrac{3\pi}{4}\right)$

9. $\csc\left(\dfrac{-3\pi}{4}\right)$

10. $\sin\left(\dfrac{3\pi}{2}\right)$

Determine the range of each function given in exercises 11 – 14.

11. $f(x) = -2\sin(4x) + 3$

12. $g(x) = 6\cos(2x) - 8$

13. $h(x) = -\sin(x) - 1$

14. $f(\theta) = 4\sin(\theta - \pi)$

In exercises 15 – 18, use the unit circle to help you answer the given question.

15. Find the ordered pair for the point on the unit circle associated with $\theta = \frac{5\pi}{4}$

16. Sketch the a unit circle and the angle represented by $\theta = \frac{7\pi}{6}$. Find the ordered pair where this line intersects the unit circle and label this point on your sketch.

17. Sketch the a unit circle and the angle represented by $\theta = -\frac{2\pi}{3}$. Find the ordered pair where this line intersects the unit circle and label this point on your sketch.

18. Find the equation of the line that intersects the unit circle at $\theta = \pi$ and at $\theta = \frac{\pi}{3}$. Answer in slope intercept form.

3.5   Trigonometric Functions and Triangles

In this section, we will further discuss trigonometric functions. We will examine the relationship between trigonometric functions and right triangles, and examine some more properties of trigonometric functions.

Right Triangles

In the last section, we focused on the relationships between the trigonometric functions and the unit circle. Now, we will examine the relationship of these functions and the unit circle with right triangles. In Figure 3.3, we start by drawing a unit circle with an angle θ and marking the corresponding coordinates on the circle. If we drop down from these coordinates to the x axis, we can form a right triangle.

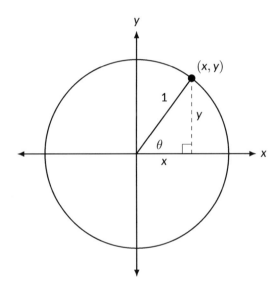

Figure 3.3: Right triangle inside of the unit circle

This triangle has a hypotenuse of 1 because the hypotenuse is the same length as the radius of the unit circle, and side lengths of $x = \cos(\theta)$ and $y = \sin(\theta)$. If we apply the Pythagorean Theorem to this triangle, we discover an interesting identity:

Notes:

$$a^2 + b^2 = c^2$$
$$(x)^2 + (y)^2 = (1)^2$$
$$(\cos(\theta))^2 + (\sin(\theta))^2 = 1^2 \qquad (3.3)$$
$$\cos^2(\theta) + \sin^2(\theta) = 1$$

There is nothing special about the choice of θ shown in Figure 3.3; this identity is true for all inputs. Notice that the input for cosine and the input for sine are the same; if the inputs are different, we cannot guarantee that the sum will be equal to 1.

This right triangle also gives us a different way of evaluating trigonometric functions, in general. With the unit circle, we saw that $\cos(\theta)$ is the x coordinate and $\sin(\theta)$ is the y coordinate, and our right triangle has a hypotenuse of 1. If we scale the triangle, the side lengths will also scale, but the size of the angles will remain the same, so the values of $\cos(\theta)$ and $\sin(\theta)$ should also remain the same. In order for this to be true, we can't just say that cosine is the length of the adjacent side and sine is the length of the opposite; instead, we will need to divide both by the length of the hypotenuse to adjust for the scaling (see Figure 3.4 for a visual explanation of the opposite and adjacent sides). This gives us the following identities:

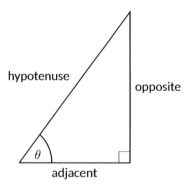

Figure 3.4: Using a right triangle to evaluate trigonometric functions

1. $\sin(\theta) = \frac{\text{opposite}}{\text{hypotenuse}}$

2. $\cos(\theta) = \frac{\text{adjacent}}{\text{hypotenuse}}$

3. $\tan(\theta) = \frac{\text{opposite}}{\text{adjacent}}$

4. $\csc(\theta) = \frac{\text{hypotenuse}}{\text{opposite}}$

5. $\sec(\theta) = \frac{\text{hypotenuse}}{\text{adjacent}}$

6. $\cot(\theta) = \frac{\text{adjacent}}{\text{opposite}}$

Notes:

Many people summarize the first three of these with SOH-CAH-TOA to help remember the identities. SOH-CAH-TOA stands for Sine is Opposite over Hypotenuse; Cosine is Adjacent over Hypotenuse, and Tangent is Opposite over Adjacent. The remaining three identities can then be formed from the definitions of cosecant, secant, and cotangent. Let's take a look at how we can use right triangles to help us evaluate our trigonometric functions.

Example 83   **Using a Right Triangle**

Suppose that $\cos(\theta) = \frac{12}{13}$. Determine all possible values of $\sin(\theta)$.

Solution   To help find the possible values of $\sin(\theta)$, we will draw a right triangle and label it using the values we already know. We know that $\cos(\theta) = \frac{12}{13}$, so we can use 12 as the length of the adjacent side and 13 as the length of the hypotenuse:

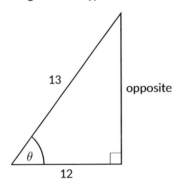

Now, we can use the Pythagorean Theorem to find the missing side length:

$$a^2 + b^2 = c^2$$
$$(12)^2 + b^2 = 13^2$$
$$144 + b^2 = 169$$
$$b^2 = 25$$
$$b = \pm 5$$

Now, we can update our drawing:

Notes:

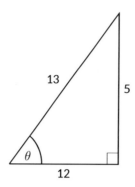

In our drawing, we labeled all of the sides with positive values because when we measure the side of a triangle we will get a positive length. From this triangle, we get $\sin(\theta) = \frac{\text{opposite}}{\text{hypotenuse}} = \frac{5}{13}$. However, this is not the only possible value of $\sin(\theta)$. We do not know the true value of θ and in our drawing assumed that it is between 0 and $\frac{\pi}{2}$. In reality, it could also be between $\frac{3\pi}{2}$ and 2π. This would mean that $\sin(\theta)$ could also have a negative value.

> The possible values of $\sin(\theta)$ are $\dfrac{5}{13}$ and $-\dfrac{5}{13}$

Inverse Trigonometric Functions

Often, we will have information about the side lengths of the triangle, but will want to know the value of the angle. This is where we will need the inverse trigonometric functions. Each trigonometric function has an inverse, but the inverses of sine, cosine, and tangent are the most commonly used. The notation for the functions is a bit tricky. We've seen that we can write $(\sin(\theta))^2$ as $\sin^2(\theta)$, however, the notation $\sin^{-1}(\theta)$ is often used to represent the inverse sine function rather than the function $\frac{1}{\sin(\theta)}$. In this book, we will instead use the notation $\arcsin(x)$ to represent the inverse sine function. This eliminates confusion over notation, but you should be aware that not all references do this. Similarly, we use $\arccos(x)$ for the inverse cosine function and $\arctan(x)$ for the inverse tangent function. These can be referred to as arcsine, arccosine, and arctangent in writing.

Notice that for each of these inverse trigonometric functions, we used x as our input rather than θ. This is because we are no longer inputting an angle, but rather a ratio of lengths. For these functions, our output will be an angle.

Notes:

Remember, when we say that two functions are inverses, we mean that there is a relationship like the following: $\arccos(\cos(\theta)) = \theta$ and $\cos(\arccos(x)) = x$. Another way of expressing this relationship is to say that if $\cos(\theta) = x$, then $\arccos(x) = \theta$. However, this is not exactly true here. When we look at trigonometric functions, we know that there are lots of angles that all result in the same value for sine, lots of angles that result in the same value for cosine, and lots of angles that result in the same value for tangent. Because we only want one output for each input, the inverse trigonometric functions use restricted outputs. Arccosine is restricted to output values between 0 and π, meaning that its range is $[0, \pi]$. This works because every possible output of cosine shows up once for angles from 0 to π. Arcsine and arctangent are restricted to output values between $-\frac{\pi}{2}$ and $\frac{\pi}{2}$, meaning that they each have a range of $\left[-\frac{\pi}{2}, \frac{\pi}{2}\right]$. For both tangent and sine, every possible output value appears once for angles between $-\frac{\pi}{2}$ and $\frac{\pi}{2}$. By restricting the ranges, we make sure these functions are well-defined, meaning they only produce one output for each input.

In practice, we can still use the unit circle to help evaluate inverse trigonometric functions. For example, if we want to evaluate $\arcsin\left(\frac{1}{2}\right)$, we will want to look at the unit circle to see where $\sin(\theta) = \frac{1}{2}$. We get two angles: $\frac{\pi}{6}$ and $\frac{5\pi}{6}$. Since the range of arcsine is restricted to $\left[-\frac{\pi}{2}, \frac{\pi}{2}\right]$, we say that $\arcsin\left(\frac{1}{2}\right) = \frac{\pi}{6}$. Similarly, we would say that $\arccos\left(\frac{1}{2}\right) = \frac{\pi}{3}$, and $\arctan(1) = \frac{\pi}{4}$.

Notes:

Exercises 3.5

Terms and Concepts

1. Sketch a right triangle that can be associated with $\theta = -\frac{\pi}{6}$ and evaluate $\sin(\theta)$.

2. Explain why the range of arcsin (x) is restricted to $\left[-\frac{\pi}{2}, \frac{\pi}{2}\right]$.

3. Explain how the unit circle helps evaluate inverse trigonometric functions.

4. True or false: $\sin^2(5x) + \cos^2(5x) = 5$. If false, correct the statement.

Problems

Evaluate each statement given in exercises 5 – 9.

5. arcsin $\left(\frac{\sqrt{3}}{2}\right)$

6. arccos $\left(\frac{-\sqrt{3}}{2}\right)$

7. arctan (-1)

8. arctan $(\sqrt{3})$

9. arcsin (0)

Use your knowledge of the trigonometric functions and their relationships to right triangles to answer the questions in exercises 10 – 19.

10. Consider a right triangle with a hypotenuse of length 5 inches. If one of the sides measures 3 inches, what is the tangent of the angle that is opposite of that side?

11. Consider a right triangle with a hypotenuse of length 5 inches. If one of the sides measures 2 inches, what is the sine of the angle that is opposite that side?

12. Imagine a circle with a radius of 2 units centered around the origin. What are the angles associated with the intersection(s) of this circle and the line $x = 1$?

13. Sketch a right triangle that can be associated with $\theta = -\frac{2\pi}{3}$ and evaluate $\tan(\theta)$.

14. Sketch a right triangle that can be associated with $\theta = \frac{5\pi}{4}$ and evaluate $\cot(\theta)$.

15. Sketch a right triangle that can be associated with $\theta = -\frac{\pi}{6}$ and evaluate $\sin(\theta)$.

16. Suppose that $\sin(\theta) = \frac{3}{5}$. What are all possible values of $\cos(\theta)$?

17. Suppose that $\cos(\theta) = \frac{5}{13}$. What are all possible values of $\tan(\theta)$?

18. Suppose that $\tan(\theta) = -1$. What are all possible values of $\csc(\theta)$?

19. A classic calculus problem involves a ladder leaning against a wall. The base of the ladder starts sliding away from the wall causing the top of the ladder to slide down the wall. If you know that the ladder has a length of 13 feet, find the value of $\cos(\theta)$ when the base of the ladder is 12 feet away from the wall.

A: Solutions To Problems

Chapter 1

Section 1.1

1. Answers will vary.
2. Multiplication
3. Answer will vary.
4. F; you should complete them from left to right
5. T; there are no parentheses, so you first square 2 and then make the answer negative..
6. 4
7. $\dfrac{7}{5}$
8. $\dfrac{5}{18}$
9. -4
10. $\dfrac{729}{8}$
11. -8
12. 1
13. 8
14. 2
15. 4
16. 5 miles
17. Re decreases.
18. 54 ft^3
19. The velocity is tripled.
20. The capacitance increases.
21. 32 ft^3
22. The final answer is not correct. There are two errors: in the second step, $-(-1)^2$ should give -1, not $+1$; in the third step, the division should have been done before the multiplication.
23. The final answer is correct.
24. The final answer is not correct. In the square root, 64 and 36 should be added before the square root is taken.
25. The final answer is not correct. In the fifth step, $(-2)(-6)$ should be 12, not -12.

Section 1.2

1. It tells us that t is the input for the function f.
2. True.
3. False.
4. True.
5. Quotient of $f(x)$ and $h(x)$; $\dfrac{f(x)}{h(x)}$
6. Composition of $h(x)$ with $g(x)$; $h(g(x))$
7. Composition of $g(x)$ with $h(x)$; $g(h(x))$
8. Product of a scalar multiple of $f(x)$ with $g(x)$; $(2f(x))(g(x))$
9. The input variable is A. The parameters are k, ε_0, and d. This is a monomial of degree 1.
10. The input variable is t. The only parameter is v_0. This is a polynomial of degree 1.
11. The input variable is t. The parameters are P, r, and n. This is an exponential function.
12. $90a^2 - 4 + b$
13. $y^2 + 2yh + h^2 - 18 + 3b$
14. $5 - y - h + 3y^4 - 3p$
15. 40
16. $x^2 + 2xh + h^2 - 12$
17. $10y - 20$
18. 2
19. 4
20. $2x + h$
21. $8t + 4h$

Section 1.3

1. Answers will vary.
2. The factors are $x - 2$, $x - 5$, and $x + 1$.
3. It means it has no real number roots and that it cannot be factored.
4. It can have a maximum of 6 linear factors since it is a sixth degree polynomial.
5. It can have a maximum of 6 roots since it is a sixth degree polynomial.
6. $6a^2b - 12ab^2 + 15a^2 - 30ab$
7. $4t^2 + 28t + 49$
8. $4x^3 + 18x^2 + 34x + 24$
9. $-3t^2 - 2t + 6$
10. $-4x^3 - 3x^2 - 5$
11. $8x^3 + 18x^2 - 14x - 15$
12. $60^5 + 480^3 + 960$
13. $g(x) = (2x + 1)^2$
14. $y(z) = (z - 5)(z - 2)$
15. $f(k) = k(k - 3)(k^2 + 3k + 9)$
16. $\theta(\gamma) = (3\gamma - 2)(2\gamma + 1)$
17. $x(z) = 3z(z + 4)(z - 2)$
18. $y(x) = (x + 2)(x^2 - 2x + 4)$
19. $f(x) = (2x + 1)(x - 2)(x + 1)$
20. $f(y) = (y - 3)(y - 4)(y + 2)$
21. $3t^2 + 3th + h^2$
22. $4x + 2h$
23. $3x^2 + 3xh + h^2 + 2x + h - 1$
24. $8x + 4h^2 + 2$
25. $x = 0$ and $x = -\dfrac{1}{2}$
26. $x = 0$, $x = \dfrac{-1+\sqrt{5}}{2}$, and $x = \dfrac{-1-\sqrt{5}}{2}$
27. $x = \dfrac{\sqrt{5}}{2}$ and $x = -\dfrac{\sqrt{5}}{2}$
28. $g(x) = 2x(2x + 1)$; the factor x pairs with the root $x = 0$ and the factor $2x + 1$ pairs with the root $x = -\dfrac{1}{2}$

29. $f(x) = x(x + \frac{1-\sqrt{5}}{2})(x + \frac{1+\sqrt{5}}{2})$; the factor x pairs with the root $x = 0$, the factor $x + \frac{1-\sqrt{5}}{2}$ pairs with the root $x = \frac{-1+\sqrt{5}}{2}$, and the factor $x + \frac{1-\sqrt{5}}{2}$ pairs with the root $x = \frac{-1-\sqrt{5}}{2}$

30. $y(x) = 4(x - \frac{\sqrt{5}}{2})(x + \frac{\sqrt{5}}{2})$; the factor $x - \frac{\sqrt{5}}{2}$ pairs with the root $x = \frac{\sqrt{5}}{2}$ and the factor $x + \frac{\sqrt{5}}{2}$ pairs with the root $x = -\frac{\sqrt{5}}{2}$

Section 1.4

1. Yes, a root function is just a power function with a fractional exponent.

2. A positive exponent means we are multiplying that term repeatedly, a negative exponent means we are dividing by that term repeatedly.

3. Exponential form

4. Radical form

5. $\dfrac{1}{\sqrt[3]{8x_1 - 5x_2 + 11}}$

6. $\dfrac{1}{\sqrt[5]{-2x + y}}$

7. $\sqrt[4]{5x - 2}$

8. $x + 2 + \dfrac{1}{x}$

9. $x + \dfrac{1}{x}$

10. $x + 3x^{2/3} + 3x^{1/3} + 1$

11. $\dfrac{25y^5}{x}, y \neq 0$

12. $64x^{16}y^9; x, y \neq 0$

13. $\dfrac{-27t}{64s^7}, t \neq 0$

14. $\dfrac{-6}{(x+2)^7}$

15. $\dfrac{4x^{1/3}}{3}$

16. e^{3x+6}

17. $e^{x^2 - x - 2}$

18. $(4x - 1)(3x + 2)^{-2/3}$

19. $e^{\theta - 2}y^2, y \neq 0$

20. $xy^4, y \neq 0$

Section 1.5

1. For the same base, they are inverses of each other.

2. 25 is 5 raised to what power?

3. e

4. Logarithms help solve exponential statements because logarithms and exponentials are inverse functions.

5. 4

6. 5.7

7. $\dfrac{1}{x}$

8. 16

9. $\log_3\left(\dfrac{16x^4}{y^2}\right)$

10. $\ln\left(8x^{2/3}y^3\right)$

11. $\log_2\left(5(3^{2x})\right)$

12. $\ln\left(\dfrac{y}{x}\right)$

13. $x = 2$

14. Not possible; we cannot raise a positive number to a power and get a negative number

15. Not possible; we cannot raise a positive number to a power and get a zero

16. $x = -1$

17. $x = 0$

18. $x = 6 + \log_3(2)$

19. $x = \dfrac{5 + \log_4(3)}{2}$

20. $x = \dfrac{-4}{5}$

21. $x = \log_6(2) - \pi$

22. $x = 0$

23. $x = \dfrac{1}{3}e^{11/4}$

24. $x = -\dfrac{\ln(3)}{\ln(2) - \ln(3)}$

25. $x = \dfrac{e^2 - 5}{2}$

Chapter 2

Section 2.1

1. Answers will vary

2. A point and a slope, or two points

3. $m = 0$

4. This is the y-intercept because $x = 0$ so it is where the line crosses the y-axis.

5. $m_2 = 2$

6. $m_2 = \frac{1}{4}$

7. $y - 2 = -2(x - 1)$

8. $y = -x + 3$

9. $y - 4 = -2(x - 0)$

10. $y = \frac{3}{2}x + \frac{3}{2}$

11. $y = \frac{3}{2}x + 4$

12. -7.8

13. $y = 3x + 8$

14. $y - 4 = \frac{1}{2}(x - 7)$, or $y - 6 = -\frac{1}{2}(x - 3)$

15. $y = -2x - 4$

16. $y - 2 = 6(x - 0)$

17. $m = -4$

18. $(2, 0)$

19. $(0, -8)$

20. The line $y = 5x + 10$ has a steeper slope.

Section 2.2

1. A strict inequality means we have $>$ or $<$.

2. We have break points when the equality statement is true or where the statement is undefined.

3. No, the statement $x^2 > 0$ is always true, but has a break point at $x = 0$.

4. We need to move everything to one side, and then we can factor or use the quadratic formula to find the roots.

5. $x \in [-3, 10]$

6. $x \in (2, \infty)$

7. $x \in [-5, 2)$

8. no values of x satisfy this statement

9. $x \in [-5, \infty)$

10. $x \in (-6, 4]$

11. $x \in (-\infty, -2) \cup (4, \infty)$

12. $3 \le x < 4$ or $x > 4$

13. $-2 \le x < 4$

14. $5 < x \le 6$ or $7 \le x < 8$

15. $x \in [2, 4)$

16. $x \in [1, 3]$

17. $x \in (-\infty, -5) \cup (3, \infty)$

18. $x \in \left[\dfrac{7 - \sqrt{89}}{2}, \dfrac{7 + \sqrt{89}}{2} \right]$

19. $x \in (-\infty, 2) \cup [7, \infty)$

20. $x \in [-5, 5]$

21. $x \in [-5, 2)$

22. $x \in \left(\dfrac{-2}{3}, \infty \right)$

23. $\theta \in [2, 3]$

24. $y \in (-4, 0) \cup (1, \infty)$

25. $x \in (-\infty, 1]$

26. $x \in (-\infty, -4) \cup [-2, -1] \cup (4, \infty)$

Section 2.3

1. It means that 2 is a valid input for the function f.

2. It means that 4 is not a valid input for the function f.

3. False; it depends on both the domain of $f(x)$ and the domain of $g(x)$.

4. False; if $g(x) = 0$, $\dfrac{f(x)}{g(x)}$ is not defined, but $f(x)$ may not be defined everywhere

5. $x \in (-\infty, -9) \cup (-9, 3]$

6. $x \in [-11, 11) \cup (11, \infty)$

7. $x \in (6, 13) \cup (13, \infty)$

8. $t \in (5, \infty)$

9. $x \in (-3, \infty)$

10. $\theta \in (-\infty, \infty)$

11. D: $(4, \infty)$

12. D: $\left(\dfrac{1}{2}, 1 \right) \cup (1, \infty))$

13. D: $(-2, 2)$

14. D: $(-\infty, -2) \cup (2, \infty)$

15. D: $(-\infty, -5] \cup [-1, \infty)$

16. D: $(-\infty, \infty)$

Section 2.4

1. Answers will vary, but should include the idea that changes on the inside are changes to the inputs, which are on the horizontal axis

2. Answers will vary, but should include the idea that applying changes on the outside of the function affects the outputs, which are the y values (height of the graph)

3. The vertical stretch

4. The horizontal shift

5. The base function is 3^t and it is being shifted 4 units to the left

6.

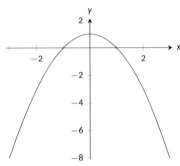

7.

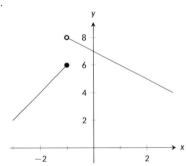

8.

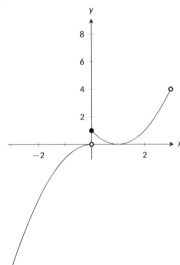

9.

A.3

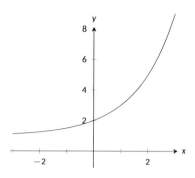

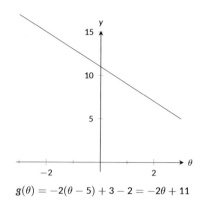

$$g(\theta) = -2(\theta - 5) + 3 - 2 = -2\theta + 11$$

10.

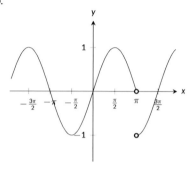

14.

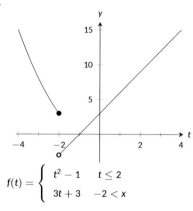

$$f(t) = \begin{cases} t^2 - 1 & t \le 2 \\ 3t + 3 & -2 < x \end{cases}$$

11.

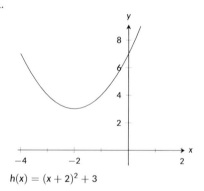

$$h(x) = (x + 2)^2 + 3$$

15.

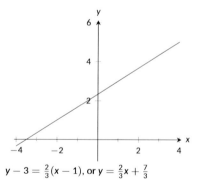

$$y - 3 = \frac{2}{3}(x - 1), \text{ or } y = \frac{2}{3}x + \frac{7}{3}$$

12. Answers will vary, but an example is

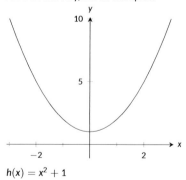

$$h(x) = x^2 + 1$$

16. $b(x) = (x + 2)^3;$

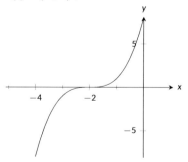

13.

17. $y(t) = (t - 3)^2;$

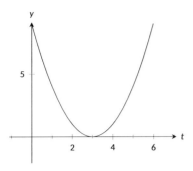

18. $f(x) = (x + 2)^2$;

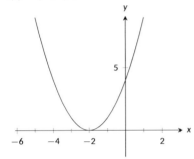

19. $x^2 - 3x + 3$; $\cos(\theta) + 3$; $3^w - w^3 + 3$

20. $(x - 2)^2 - 3(x - 2)$; $\cos(\theta - 2)$; $3^{w-2} - (w - 2)^3$

21. $(x + 1)^2 - 3(x + 1) - 2$; $\cos(\theta + 1) - 2$; $3^{w+1} - (w + 1)^3 - 2$

22. $(x - e)^2 - 3(x - e) - \pi$; $\cos(\theta - e) - \pi$; $3^{w-e} - (w - e)^3 - \pi$

23. $(-x)^2 - 3(-x)$; $\cos(-\theta)$; $3^{-w} - (-w)^3$

24. $-(x^2 - 3x)$; $-\cos(\theta)$; $-(3^{-w} - w^3)$

25. (a) no; it is not symmetric about the y-axis

(b) yes; it has rotational symmetry

(c) no; it does not have a horizontal asymptote

(d) no; it does not have a periodic (repeating) pattern

26. (a) no; it is not symmetric about the y-axis

(b) no; it does not have rotational symmetry

(c) yes; it does not have a horizontal asymptote on one side and grows without bound on the other

(d) no; it does not have a periodic (repeating) pattern

Section 2.5

1. After completing the square, you can quickly identify the horizontal and vertical shifts

2. $x = 2$ is not a root of $f(x)$ because $f(2) = 3$, not 0.

3. It represents a horizontal stretch/shrink because it is on the inside of the function.

4. $a = 2$ and $b = -6$

5. $f(x) = (x - 2)^2 + 2$; $a = -2$; $b = 2$

6. $g(x) = (x + 10)^2 - 60$; $a = 10$; $b = -60$

7. $h(x) = (x - 4)^2 - 11$; $a = -4$; $b = -11$

8. $m(x) = (x - 11)^2 - 125$; $a = -11$; $b = -125$

9. $n(x) = (x - 3)^2 - 11$; $a = -3$; $b = -11$

10. $p(x) = (x + \frac{11}{2})^2 - \frac{105}{4}$; $a = \frac{11}{2}$; $b = -\frac{105}{4}$

11. $p(x) = (x + \frac{13}{2})^2 - \frac{169}{4}$; $a = \frac{13}{2}$; $b = -\frac{169}{4}$

12. $f(x) = 9(x - \frac{2}{3})^2 + 8$; $a = -\frac{2}{3}$; $b = 8$; $c = 9$

13. $f(x) = (x - 1)^2 + 1$; $a = -1$; $b = 1$; $c = 1$

14. $h(x) = 4(x - \frac{1}{2})^2 - 5$; $a = -\frac{1}{2}$; $b = -5$; $c = 4$

15. $w(x) = 4(x + \frac{1}{2})^2 + 5$; $a = \frac{1}{2}$; $b = 5$; $c = 4$

16. $y(x) = 9(x + 1)^2 - 5$; $a = 1$; $b = -5$; $c = 9$

17. $f(t) = (t + 1)^2 + 2$;

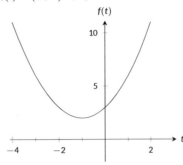

18. $p(q) = (q - \frac{1}{3})^2 - \frac{1}{9}$;

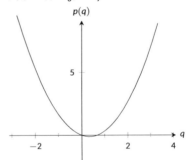

19. $y(x) = (x + 2)^2 - 2$;

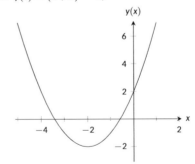

20. $f(x) = (x - 2)^2 + 2$;

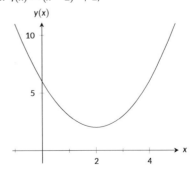

21. $f(x) = x^2 - 2x - 1$

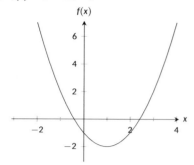

22. $g(x) = -x^2 - 6x - 5$

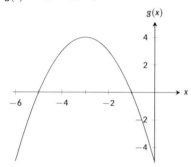

23. $h(x) = x^2 - 6x + 13$

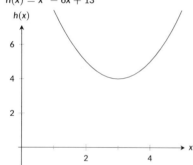

24. $x(y) = y^2 + 4y + 3$

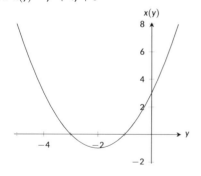

Chapter 3

Section 3.1

1. Yes, but it can be quite difficult, especially if it has many parameters

2. 2 distinct real roots; 1 repeated real root; a complex conjugate pair of roots

3. 7

4. F; it must have at least one real solution since complex solutions come in pairs

5. cubic

6. linear

7. quadratic

8. trigonometric

9. quartic, or a statement of degree 4

10. cubic

11. It is possible to solve; $x = \dfrac{5y}{y^2 - y + 3}$

12. $\dfrac{x + 5 \pm \sqrt{-11x^2 + 10x + 25}}{2x}$

13. It is possible to solve; $q = \dfrac{3t^2 - 2m^3}{8t + 5m}$

14. It is possible to solve;
$$a = \frac{-(3bc^2) \pm \sqrt{(3bc^2)^2 - 4(2bc^3 + 4c^2)(-3b - 4c)}}{2(2bc^3 + 4c^2)}$$

15. It is possible to solve; $b = \dfrac{4c - 4a^2c^2}{2a^2c^3 + 3ac^2 - 3}$

16. Not possible to solve for x; it is inside of a logarithm and has a linear term

17. It is possible to solve; $y = 2^{x + e^z - \log_2(x)}$ or $y = \dfrac{2^{x + e^z}}{x}$

18. It is possible to solve; $z = \ln\left[\log_2(xy) - x\right]$

19. $x = -4, 4$

20. $x = -4i, 4i$

21. $x = 2 + \sqrt{13}, 2 - \sqrt{13}$

22. $x = 1 + 2i, 1 - 2i$

23. $x = \dfrac{-1 + 2i}{5}, \dfrac{-1 - 2i}{5}$

24. $x = 2$

25. $x = -2, -1, 2$

26. $x = 2 - \sqrt{3}, 2 + \sqrt{3}$

27. $x = -1, \dfrac{-3 + \sqrt{5}}{2}, \dfrac{-3 - \sqrt{5}}{2}$

28. Two real solutions

29. A complex conjugate pair

30. Two real solutions

31. One repeated solution

32. Three real solutions

Section 3.2

1. When one or both functions are defined implicitly

2. Explicitly; y is isolated

3. Implicitly; y is not isolated

4. Answers will vary, but graphing helps you determine how many intersections points exist, but does not always clearly show the exact values

5. 1

6. 2

7. 2

8. 3

9. 1

10. 0

11. 0

12. 1

13. $(0, -1)$ and $(1, 0)$

14. $(\frac{4}{5}, \frac{3}{5})$ and $(-\frac{4}{5}, -\frac{3}{5})$

15. $(0, 1)$ and $(3, 4)$

16. $(1, 0)$ and $(2, 0)$

17. $(\frac{3+\sqrt{21}}{2}, 5)$ and $(\frac{3-\sqrt{21}}{2}, 5)$

18. $(1, 3)$, $(2, 1)$, and $(4, -3)$

19.

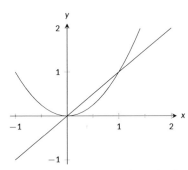

Points of intersection are $(0, 0)$ and $(1, 1)$

20.

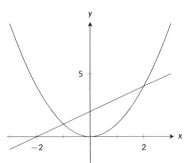

Points of intersection are $(-1, 1)$ and $(2, 4)$

21.

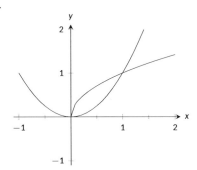

Points of intersection are $(0, 0)$ and $(1, 1)$

22.

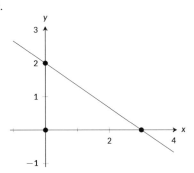

Points of intersection are $(0, 0)$, $(0, 2)$, and $(3, 0)$

Section 3.3

1. No, the numerator and denominator have no common factors

2. Yes, $x + 3$ is repeated twice

3. A quadratic that has no real valued roots

4. Answers will vary; $x^2 + a$ is an example if $a > 0$

5. $\dfrac{-5}{36}$

6. $\dfrac{x^2 - b^2}{xb}$

7. $\dfrac{x^2 + xy - xy^2}{xy^2 + y^3}$

8. $\dfrac{-2x}{-x^2 + 4x - 1}, x \neq 0$

9. $\dfrac{1}{x^2 - bx}, b \neq 0$

10. $\dfrac{A}{x+2} + \dfrac{B}{(x+2)^2} + \dfrac{C}{(x+2)^3}$

11. $\dfrac{A}{s-1} + \dfrac{B}{(s-1)^2} + \dfrac{C}{2s-5} + \dfrac{D}{s+3}$

12. $\dfrac{A}{t+1} + \dfrac{B}{(t+1)^2} + \dfrac{C}{(t+1)^3} + \dfrac{Dt+E}{t^2+1} + \dfrac{Ft+G}{(t^2+1)^2}$

13. $\dfrac{A}{x-4} + \dfrac{Bx+C}{x^2+x+5}$

14. $\dfrac{A}{x+1} + \dfrac{B}{x-1} + \dfrac{Cx+D}{x^2+1}$

15. $\dfrac{A}{s+1} + \dfrac{Bs+C}{s^2-s+1}$

16. $\dfrac{A}{t-5} + \dfrac{B}{t-1}$

17. $\dfrac{-1}{x+2} + \dfrac{2}{x-1}$

18. $\dfrac{1/(2a)}{x-a} - \dfrac{1/(2a)}{x+a}$

19. $\dfrac{1}{s} + \dfrac{s-1}{s^2+4}$

20. $\dfrac{3}{y+2} - \dfrac{2}{y+1}$

21. $\dfrac{-1}{x+1} + \dfrac{1}{x-1} + \dfrac{2}{(x-1)^2}$

22. $\dfrac{1/2}{x} + \dfrac{1/5}{2x-1} - \dfrac{1/10}{x+2}$

Section 3.4

1. Tangent is undefined whenever cosine is 0.

2. Cosecant is undefined whenever sine is 0.

3. The range describes the possible output values of the function.

4. The x coordinate tells you the value of cosine for that angle and the y coordinate tells you the value of sine for that angle.

5. See Figure 3.2.

6. 1

7. $\dfrac{\sqrt{2}}{2}$

8. $\dfrac{\sqrt{2}}{2}$

9. $-\sqrt{2}$

10. -1

11. $[1, 5]$

12. $[-14, -2]$

13. $[-2, 0]$

14. $[-4, 4]$

15. $\left(-\frac{\sqrt{2}}{2}, -\frac{\sqrt{2}}{2}\right)$

16.

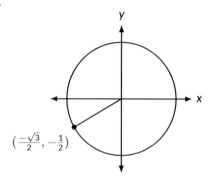

$\left(\frac{-\sqrt{3}}{2}, -\frac{1}{2}\right)$

17.

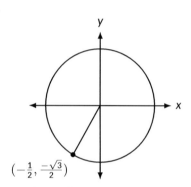

$\left(-\frac{1}{2}, \frac{-\sqrt{3}}{2}\right)$

18. $y = \frac{\sqrt{3}}{3}x + \frac{\sqrt{3}}{3}$

Section 3.5

1.

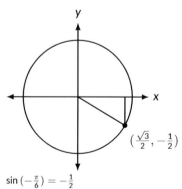

$\sin\left(-\frac{\pi}{6}\right) = -\frac{1}{2}$

2. The range of arcsin (x) is restricted so that every input only has one output.

3. You can look through the coordinates to see when the regular function is equal to the input of the inverse function.

4. False, $\sin^2(5x) + \cos^2(5x) = 1$.

5. $\frac{\pi}{3}$

6. $\frac{5\pi}{6}$

7. $\frac{-\pi}{4}$

8. $\frac{\pi}{3}$

9. 0

10. $\pm\frac{3}{4}$

11. $\frac{2}{5}$

12. $\theta = \frac{\pi}{3}, \frac{5\pi}{3}$

13.

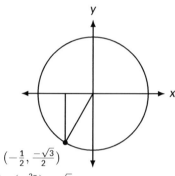

$\left(-\frac{1}{2}, \frac{-\sqrt{3}}{2}\right)$

$\tan\left(-\frac{2\pi}{3}\right) = \sqrt{3}$

14.

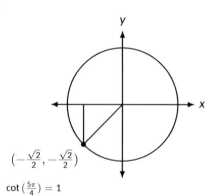

$\left(-\frac{\sqrt{2}}{2}, -\frac{\sqrt{2}}{2}\right)$

$\cot\left(\frac{5\pi}{4}\right) = 1$

15.

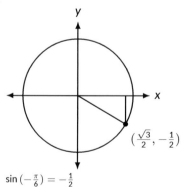

$\left(\frac{\sqrt{3}}{2}, -\frac{1}{2}\right)$

$\sin\left(-\frac{\pi}{6}\right) = -\frac{1}{2}$

16. $\frac{4}{5}$ and $-\frac{4}{5}$.

17. $\frac{12}{5}$ and $-\frac{12}{5}$.

18. $\sqrt{2}$ and $-\sqrt{2}$

19. $\frac{12}{13}$ if θ is the angle between the ladder and the floor; $\frac{5}{13}$ if θ is the angle between the ladder and the wall

Made in the USA
Lexington, KY
07 September 2019